A
SYNOPTIC CLASSIFICAT
LIVING ORGANISM

A SYNOPTIC CLASSIFICATION OF LIVING ORGANISMS

EDITED BY
R. S. K. BARNES
BSc, PhD
St Catharine's College,
Cambridge, UK

BLACKWELL SCIENTIFIC PUBLICATIONS
OXFORD LONDON EDINBURGH
BOSTON MELBOURNE

Editorial offices:
Osney Mead, Oxford, OX2 OEL
8 John Street, London, WC1N 2ES
9 Forrest Road, Edinburgh, EH1 2QH
52 Beacon Street, Boston
Massachusetts 02108, USA
99 Barry Street, Carlton
Victoria 3053, Australia

First published 1984

Set by Macmillan (India) Ltd
and printed and bound
in Great Britain by
Billing & Sons Ltd
Worcester

Distributed in the USA by
Sinauer Associates, Inc.,
Publishers, Sunderland,
Massachusetts

British Library
Cataloguing in Publication Data

A synpotic classification of living organisms.
1. Biology—Classification
I. Barnes, R. S. K.
574'.012 QY83

ISBN 0 632 01145 9

IN MEMORY
OF THE
RHINOGRADENTIA
AND THE
HOOAKHA-HUTCHI

Contents

Contributors, viii

Preface, ix

Introduction, 1

Kingdom Monera, 7

Kingdom Protista, 25

Kingdom Fungi, 89

Kingdom Plantae, 111

Kingdom Animalia, 129

References and Further Reading, 259

Index to Taxa, 261

Contributors

DR R. S. K. BARNES *St Catharine's College and Department of Zoology, University of Cambridge*
Monera, Plantae and Animalia

PROFESSOR J. D. DODGE *Department of Botany, Royal Holloway College, University of London*
Protista, part

DR H. J. HUDSON *Fitzwilliam College and Botany School, University of Cambridge*
Fungi

DR D. J. PATTERSON *Department of Zoology, University of Bristol*
Protista, part

PROFESSOR M. A. SLEIGH *Department of Biology, University of Southampton*
Protista, part

Preface

Many biologists come across the names of groups of organisms with which they are unfamiliar or of which they have never heard, and wish to know something about them. Existing works to which they can turn either cover only a limited range of groups (often in great detail), or cover all types but in only the most general of fashions, or are massive and correspondingly priced tomes found in only a few libraries. We have tried to remedy this situation.

We hope that this book will serve various functions. It is as comprehensive, concise and small (and therefore inexpensive) as we can make it, and therefore we would like to think that it will function as the sort of dictionary/mini-encyclopaedia of organismal classification and diversity that all professional, amateur and student biologists could have on their desks. It may even be useful to a departmental typist in establishing the spelling appropriate to some unfamiliar, barely decipherable, manuscript word. Such is the pressure of time in school and college courses these days that student biologists are taught less and less of the diversity of life: in these pages, maybe, they can find out what sort of organism a foraminiferan is and to what it is related; the differences between the major groups of photosynthetic bacteria; whether or not a gastrotrich has a body cavity; and the number of species of bryophytes. It will be apparent that we have not produced this classification for specialists on the groups themselves; we assume that they will know their own groups in much more detail than could ever be included here. We do hope, however, that bacteriologists, for example, will find the synoptic classification of animals useful, and vice versa. It is clearly not a book to sit down and read, but the index to the taxa included should permit it to be used as a taxonomic vade-mecum.

We are most grateful to Jan Parr and Sue Wickison for enlivening the text with vignettes of each of the 67 eukaryote phyla distinguished in this classification.

Cambridge, 1983 R. S. K. B.

INTRODUCTION

R. S. K. BARNES

The following pages present an outline, synoptic account of the classification of living organisms from the prokaryotic bacteria, through the protists, to the multicellular fungi, plants and animals.

Taxonomic categories of whatever level are arbitrary, man-made distinctions imposed on a continuum of natural variation: it cannot be stressed too forcefully that genera, orders, classes and the like have no objective existence or intrinsic biological meaning. Nevertheless, as a result of shared ancestry with differing degrees of subsequent independent evolution, any organism will show different degrees of similarity to other organisms and this is—and for the forseeable future will continue to be—reflected in a hierarchical classificatory scheme based upon morphological resemblance and inferred phylogenetic relationship.

In the scheme put forward here, we are endeavouring to present a consensus and are not proposing anything radically novel; we have tried to be up-to-date in the sense of reflecting modern views on inter-relations without being avant-garde; and we hope to have created something broadly useful in that we have steered a middle course between the two desiderata of providing a simple, unambiguous, working classification and of producing a system which reflects phylogenetic affinity most accurately. This is not a textbook of systematics or taxonomy, neither is it a specialist treatise on one particular group of organisms: we therefore have not burdened the casual user with series of alternative classifications together with the relevant arguments for and against each option. Because the process of pigeon-holing is arbitrary, there are often many conflicting approaches and no one scheme can ever be regarded as the 'right' one. We have adopted those which seem to us to be of maximum value to the majority of potential users whilst, at the same time, most accurately portraying likely phylogeny.

Taxonomic hierarchies contain many categories (one recent classification of mammals interposes thirteen between the levels of class and genus), however constraints of space and clarity have in general restricted us to three: phylum, class and order. Our concept of a phylum is that of a group of organisms which appear to be related one to each other but whose relationships with other such groups are debatable or entirely conjectural. Pragmatically, the essence of phylum status is ignorance of affinity, usually consequent on lack of relevant fossil material and wide morphological distance between a group in question and any other. Within a phylum, the categories of class and order represent groupings of anatomical distinctiveness, but of known or assumed ancestry.

Any attempt to construct a single, unified classification of all types of living organism must confront (at least) two major problems: different branches of biology have their own jealously guarded traditions; and some types of organism have been studied much more than have others. Thus, traditionally the highest ranking taxon in botany is the division and that in zoology is the phylum, but the two are not equivalent or interchangeable; and the concept or level of a single taxon may vary greatly between groups—the orders Coleoptera (beetles) and Decapoda (shrimps, lobsters and crabs) contain much more diverse assemblages than do the orders Tubulidentata (aardvarks) and Psittaciformes (parrots). We have tried to produce a uniform, even and consistent classification and hence where, in certain well-known groups, the concept of, for example, an order is particularly narrow (as it is in various vertebrates and angiosperms) we have replaced it by the superorder category, which in these organisms is probably more nearly equivalent to the concept of an order as used elsewhere.

In this, as in other areas, we hope our scheme is a reasonable and balanced compromise. We trust that this will not be a recipe for pleasing nobody. Collectors of statistics may like to know that the classification divides living organisms between the equivalent of 70 phyla, 208 classes and 749 groups of ordinal status. Finally, we must acknowledge our debt to all those systematists whose labours largely originated the systems and taxa adopted here.

Notes to users of this book

1. The brief descriptions or diagnoses given for each taxon are based solely on the features shown by living members of the groups concerned, even though the taxon may have been erected for—and defined formally on the basis of—fossil material differing in morphology from their living descendants.
2. Divergence within a taxon frequently results in some species lacking individual characters otherwise diagnostic of that taxon—there will always be exceptions to anatomical definitions and, therefore, the descriptions and diagnoses are to be regarded as generally, not invariably, applicable.
3. The viruses are not considered 'organisms' within the meaning of this book and are, therefore, excluded; and the composite 'lichens' are not treated as a separate taxon, the component fungal and algal symbionts being covered where appropriate under their own Kingdoms.

4. There is no consistency across the five Kingdoms in the word-endings appropriate to different taxa, although within the groups traditionally regarded as the province of botany there is uniformity of suffix below the level of class (subclass = ... idae; order = ... ales; family = ... aceae) and in zoology there is uniformity below the level of order (family = ... idae). Zoological orders often end in ... ida, but a wide variety of other endings are also used. It should be noted, therefore, that both subclasses and families referred to below may end in ... idae, and that the family ending, ... aceae should not be confused with that of a class of algal protists, ... phyceae. This potential confusion is regrettable but unavoidable until zoologists, botanists and microbiologists adopt a unified nomenclatorial system.

KINGDOM MONERA

R.S.K. BARNES

Monerans are small, prokaryote, rod-shaped (bacilli), spherical (cocci) or helically spiral cells occurring singly or, more rarely, colonially in filaments (a few form mycelia lacking cross walls) and with, between them, a very wide range of aerobic or anaerobic, chemoheterotrophic, chemoautotrophic, photoautotrophic or photoheterotrophic metabolic pathways. The individual cells are usually between 0.1 and 2 μm wide, lack intracellular organelles such as mitochondria, chloroplasts or endoplasmic reticula, and possess the enzymes associated with synthesis, oxidative reactions, etc. bound to the cell membrane or to pockets of that membrane. Synthesis of DNA continues throughout life and these nucleic acids are not organized into, or contained in, chromosomes; neither is the nuclear material enclosed within a membrane. The ribosomes are of a characteristically small size. In most, the cells possess cell walls containing the distinctive polymer peptidoglycan arranged either with teichoic acids and polysaccharides as a thick uniform monolayer (these walls stain a purple colour—positive—in the Gram test), or as a dense, thin, inner rigid layer outside of which are one or two peptidoglycan-less surface layers formed of proteins, lipopolysaccharides and lipoproteins (these stain a light pink—negative—in the Gram test).

Cells may be non-motile, may move by gliding, or may possess from one to many simple, bacterial flagella composed of flagellin and lacking the '9 + 2' microtubular organization of eukaryote flagella or cilia. Cell division is usually by direct binary fission (without mitosis) although some produce buds; multiplication is normally entirely asexual, even if sexual, genetical material is simply transferred from a donor to a recipient.

Monerans occur in all known habitat types whether oxic or anoxic, including the most acid, alkaline or saline of aquatic systems, intra- or extracellular parasitically in protists, fungi, plants and animals (many being pathogenic), and systems containing high concentrations of hydrogen sulphide; almost all organic substances (including various plastics and petroleum) and several inorganic ones can provide substrates for their growth.

The prokaryotic monerans were the basal stock from which the eukaryotic protists evolved (and via the protists, the three other Kingdoms); the eukaryotic condition being formed, one body of opinion suggests, by the permanent symbiotic association of a number of different prokaryotes. On this basis, the eukaryotic cell is a chimaera, the intracellular organelles characteristic of eukaryotes being, in origin, individual prokaryotic organisms.

Relationships between the different living monerans, however, are very poorly, if at all, understood and no suggested phylogenetic scheme has attracted general support. They can relatively easily be assigned to a number of largely pragmatic groups based on gross morphology, including the form of the cell wall and its reaction in the Gram test, the type of movement, mode of multiplication, habitat type, metabolic pathways and source of carbon and energy, etc., but these groups may or may not be natural in the sense of including related, and only related, organisms. Except for these groups, there is no general agreement on taxonomic classification above the level of genus (and so, for example, the numbers of families in the various groupings are not quoted in the following account); even some genera, such as *Micrococcus*, are probably phylogenetically heterogeneous. In the following account, the monerans have been divided into groups largely on the basis of those listed in the latest edition (8th Edn, 1974) of *Bergey's Manual of Determinative Bacteriology*, but with modifications to accommodate some of the views of Fox *et al.* (1980) and of other recent authors. Some of the groups may ultimately be regarded as distinct phyla; others may be associated together in combinations other than those presented here; and yet others may eventually be shown to be artificial: the majority are given common names rather than formal taxonomic designations.

All monerans are 'bacteria', although the Cyanobacteria were for many years known as the 'blue-green algae'. Some 5000 species have been described and many more probably await description.

The Archaebacteria

Monerans without muramic acid in their cell walls, with a distinctive type of transfer RNA, with a branched-chain ether-linked lipid as a major component of the membrane, and with a characteristic RNA polymerase subunit structure.

The Archaebacteria are regarded by some as forming a separate Kingdom, distinct from all other prokaryotes.

Methanogenic bacteria

Gram-positive or gram-negative obligate anaerobes; reduce CO_2 with H_2 to form methane; formate, acetate or methanol may also serve as hydrogen donors; non-motile or flagellate rods, cocci or spirals; multiply by binary fission; with characteristic co-enzymes (M and F_{420}); in aquatic sediments or in alimentary tracts of animals.

Halobacteria

Gram-negative aerobic or facultatively anaerobic chemoheterotrophs using amino acids as a source of carbon and energy; under anaerobic conditions, photophosphorylate using a light-driven bacteriorhodopsin proton pump; contain carotenoids in cell membrane; occur as non-motile cocci or polar-flagellated rods; require minimum environmental NaCl concentration of 2 mol/l; in salt lakes, brines and on salt-preserved products.

Thermoacidophilic bacteria

Gram-negative facultative chemoautotrophs using elemental sulphur as an energy source; occur as lobed, non-motile cocci; in hot acid freshwaters (pH 1–3; temp. 55–85°C).

The Oxyphotobacteria

Gram-negative aerobic bacteria capable of oxygenic photosynthesis using water as the electron donor and liberating oxygen; mostly obligately photoautotrophic; all possess chlorophyll *a* of the type also found in the photosynthetic eukaryotes, but not in other prokaryotes, as well as other pigments; photosynthetic pigments embedded in membranous thylakoids internal to the cell membrane. The oxyphotobacteria, therefore, have an algal-like photophysiology but a moneran anatomy.

Cyanobacteria

Large, non-motile or gliding rods or cocci occurring singly or colonially in filaments enclosed within a common sheath formed by outer layer of cell wall; with phycobiliprotein pigments in disc-shaped granules on outer surface of the single thylakoids; large thylakoids not enclosed by a membrane; filamentous forms often with large, thick-walled resting cells, and heterocysts in which molecular nitrogen is fixed under anoxic conditions; sheath of filamentous forms, gelatinous, fibrous or calcified; planktonic species often with gas vacuoles; multiply by binary or multiple fission or fragmentation of filaments; some facultatively anoxyphotosynthetic using H_2S as the electron donor; in soil (including deserts), in freshwaters (including hot springs and alkaline lakes), in the sea, and symbiotically (including with fungi in 'lichens'). 'Blue-green algae' or 'blue-green bacteria'.

Rival classifications are used by bacteriologists and algologists.

CHROOCOCCALES

Single or colonial cocci; multiply by simple binary fission.

CHAMAESIPHONALES

Cocci attached singly or in clusters to substratum; multiply by releasing exospores.

PLEUROCAPSALES

Coccoid; unicellular or united in common pseudofilaments; multiply by forming non-resistant intracellular baeocytes, parent organism then disintegrating.

NOSTOCALES

Filamentous, with or without false branching; many with heterocysts fixing molecular N_2; multiply by detaching small filaments (hormogonia).

STIGONEMATALES

With true branched filaments; mostly with heterocysts fixing molecular N_2; multiply by detaching small filaments (hormogonia).

Prochlorophytes

Non-motile, unicellular cocci resembling the chloroplasts of photosynthetic eukaryotes; with chlorophyll *b* as well as chlorophyll *a* but without phycobiliproteins; with paired or stacked thylakoids; obligate extracellular symbionts of tropical, marine colonial ascidians on or in the host's cloacal chamber. A single genus.

The Eubacteria

A large catch-all category for all those bacteria with muramic acid in the cell wall which, if they are photosynthetic, do not evolve oxygen and do not use water as the electron donor. Doubtless, many of the 38 groupings detailed below will eventually be accorded equivalent ranking with the Oxyphotobacteria.

Anoxyphotobacteria

A diverse assemblage of gram-negative photobacteria capable of anoxygenic photosynthesis using H_2S, H_2 or various organic compounds to reduce CO_2 according to the general equation

$$2(H_2X) + CO_2 \longrightarrow (CH_2O) + H_2O + 2X$$

in which X is never oxygen; they contain bacteriochlorophylls *a*, *b* or *c* and *d* or *e*; many can also fix molecular nitrogen.

GREEN SULPHUR BACTERIA

Green or brown, non-motile rods (sometimes cocci) with bacteriochlorophylls *c* and *d* or *e*; photosynthetic membranes in form of oblong Chlorobium vesicles just under cell membrane; obligately anaerobic photoautotrophs using H_2 or reduced sulphur compounds as electron donors, depositing sulphur outside cells; many fix molecular N_2; multiply by binary fission; in anaerobic aquatic sediments.

GREEN NON-SULPHUR BACTERIA

Gliding filamentous rods; facultatively photoautotrophic, photoheterotrophic or chemoheterotrophic; in absence of O_2, with bacteriochlorophyll *c* and photosynthetic membranes in form of Chlorobium vesicles, under these conditions photoheterotrophic, and in additional presence of H_2S photoautotrophic; in presence of O_2, chemoheterotrophic; when heterotrophic, require glycerol, acetate, etc.; in freshwaters, including neutral or alkaline hot springs.

PURPLE SULPHUR BACTERIA

Brownish to purple, non-motile or polarly flagellated rods, cocci or spirals with bacteriochlorophylls *a* or more rarely *b*, and with photosynthetic membranes continuous with cell membrane in form of vesicles; essentially anaerobic photoautotrophs using H_2S or other reduced sulphur compounds as electron donors, elemental sulphur accumulating transiently within cell in form of globules until further oxidized to sulphate; photoassimilation of acetate and some other organic compounds possible; multiply by binary fission; in sulphide-rich aquatic systems.

PURPLE NON-SULPHUR BACTERIA

Purple, motile, flagellated rods, ovoids or spirals with bacteriochlorophyll *a* or more rarely *b*, and with photosynthetic membranes continuous with cell membrane in form of vesicles, stacks of lamellae or tubes; facultatively photoheterotrophic, chemoheterotrophic or photoautotrophic; predominantly photoheterotrophs utilizing wide range of organic compounds; in presence of O_2 and absence of light, chemoheterotrophs using same range of organic compounds; in absence of O_2 and presence of low levels of H_2S, photoautotrophic without accumulating globules of elemental sulphur; cannot withstand high levels of H_2S; many able to fix molecular N_2; multiply by budding or binary fission; in sedimentary debris in aquatic systems.

Pseudomonads

Gram-negative, aerobic, unicellular rods moving by polar flagella and multiplying by binary fission; essentially chemoheterotrophic, respiring a wide variety of organic compounds; some can respire facultatively under anoxic conditions by reducing (denitrifying) nitrates; some facultatively chemoautotrophic using molecular H_2; many survive on inorganic substrates if provided with one or a few simple organics; free-living in soil and aquatic systems, and commensal and parasitic in plants and animals.

Aerobic nitrogen-fixing bacteria

Gram-negative, aerobic, oval or rod-shaped chemoheterotrophs, non-motile or with subpolar or peritrichous flagella, multiplying by binary fission; capable of fixing molecular N_2 in aerobic surroundings; often produce copious slime; require molybdenum; in soil and aquatic systems, on surfaces of plants, plant pathogenic, and in symbiotic association with the roots of leguminous plants (in nodules).

Facultatively anaerobic rods

Gram-negative, rod-shaped chemoheterotrophs with a respiratory metabolism under aerobic conditions and fermentative when anaerobic, some respire anaerobically by reducing (denitrifying) nitrates; carbohydrates fermented, wide range of organics respired; unicellular; non-motile or with flagella; multiply by binary fission.

ENTEROBACTERIA

Straight rods, non-motile or with peritrichous flagella; ferment glucose and other carbohydrates producing organic acids and often gas; free-living in soil and aquatic systems, symbionts in animal alimentary tracts, and plant and animal pathogens, causing, e.g., bacterial dysentery, typhoid, bubonic plague, gastroenteritis and food-poisoning.

VIBRIOS

Curved rods, mostly with single polar flagellum; some can adopt coccoid form in cultures; some produce light of 490 nm wave length continuously in presence of O_2, either as free-living cells or in symbiotic association with fish, squid, etc.; free-living, especially in aquatic systems, and symbiotic or parasitic in animals, causing, e.g., cholera.

Dissimilatory sulphate-reducers

Gram-negative, anaerobic rods with polar or peritrichous flagella; chemoheterotrophs with a respiratory metabolism reducing sulphate to sulphide or elemental sulphur and requiring lactate, pyruvate, etc.; multiply by binary fission; free-living in anaerobic aquatic sediments or symbiotic in animal alimentary tracts.

Spirillas

Gram-negative, helically-coiled, rigid-walled, aerobic chemoheterotrophs with bipolar tufts of flagella; prefer low oxygen tensions, one form facultatively chemoautotrophic using H_2; respire amino and organic acids; multiply by binary fission; free-living in freshwaters and the sea, some commensal in gut of bivalve molluscs.

Curved bacteria

Curved to ring-shaped, small, gram-negative chemoheterotrophs requiring, where known, amino and organic acids; ring-shaped forms, free-living, aerobic, in marine plankton and epiphytic, little known; curved forms (bdellovibrios), microaerophilic parasites, penetrating various gram-negative bacteria and multiplying, by multiple fission, in the host's periplasmic space, with single polar flagellum; in soil and aquatic systems.

Gram-negative anaerobic fermenters

Obligately anaerobic cocci or rods occurring as symbionts or parasites in alimentary tracts, respiratory passages and some other tissues of animals; non-motile or with flagella in various patterns; chemoheterotrophs fermenting sugars and amino acids to acetates, formates, butyrates, etc.; incapable of synthesizing porphyrins; multiply by binary fission.

Coccobacilli

Gram-negative, non-motile, aerobic chemoheterotrophs often occurring as pairs of cells; multiply by binary fission; mostly parasitic in the mucous membranes of animals, causing, e.g., gonorrhoea and meningitis.

Sheathed bacteria

Large, gram-negative, rod-shaped chemoheterotrophs forming chains of non-motile or flagellate cells enclosed in tubular sheaths and often attached to a substratum; sheaths chemically complex, with proteins, polysaccharides and lipids and, in several, with massive deposits of iron and manganese oxides on, or in, the sheaths; with aerobic respiratory metabolism utilizing sugars, alcohols and organic acids; multiply by transverse or longitudinal fission producing motile cells with polar or subpolar flagella; free-living in freshwaters.

Budding and/or prosthecate bacteria

Diverse group of gram-negative, free-living, aerobic, rod-shaped chemoheterotrophs mostly with semi-rigid extensions from the cell (prosthecae); each cell with one to several prosthecae forming attachment stalk, involved in multiplication, or of unknown function; multiplication by budding from tips of prosthecae or from cell surface, or by binary fission to produce dissimilar daughter cells (a non-motile prosthecate rod and a flagellate non-prosthecate swarmer which eventually converts into a non-motile prosthecate rod); budding may produce mycelium-like network of cells; some facultatively methylotrophic coupled with denitrification, others oxidize and deposit iron and manganese salts; many attach to other organisms or to inert substrata; mainly in aquatic systems, a few in soil.

Chemolithotrophic bacteria

Aerobic, gram-negative chemoautotrophs obtaining their carbon from CO_2 and their reducing power from various inorganic substances (NH_3, NO_2^-, reduced sulphur compounds, Fe^{2+}, CH_4, etc.); multiply by binary fission.

NITRIFYING BACTERIA

Essentially obligate chemoautotrophs (although a few can use acetate) oxidizing NH_3 to NO_2^-, or NO_2^- to NO_3^-; non-motile or flagellated rods, cocci, spirals or irregularly lobed; often with extensive intrusive cell membrane systems in form of lamellae or tubes; in soils and aquatic systems.

COLOURLESS SULPHUR BACTERIA

Facultative or obligate chemoautotrophs oxidizing H_2S, S or $S_2O_3^{2-}$ to SO_4^{2-}; non-motile or flagellated rods, cocci or spirals; with or without transient intracellular granules of sulphur; in soils and aquatic systems.

SIDEROCAPSANS

Little known rods and cocci with surrounding capsules; deposit oxides of iron or manganese on or in the capsules or on extracellular material;

assumed to be chemoautotrophic; in iron or manganese-rich soils or freshwaters.

METHYLOMONADS

Obligate methylotrophs deriving carbon and energy from oxidizing methane or methanol; non-motile or polarly flagellated rods and cocci; with extensive intrusive cell membrane systems in form of paired membranes or stacks of discs; some form resistant exospores or cysts; in aquatic sediments.

Rickettsias

Very small, obligately parasitic, gram-negative rods or cocci with limited metabolic capabilities.

RICKETTSIALES

Rods; intracellularly parasitic in some insects and arachnids without producing disease symptoms, but cause diseases, e.g., typhus and spotted fevers, if passed to a vertebrate animal; with autonomous energy-yielding metabolism; multiply by binary fission.

CHLAMYDIALES

Cocci; pass from vertebrate to vertebrate without intervening arthropod stage; metabolically extremely limited, possibly even energy parasites; cell surface with numerous, short, cylindrical processes; with distinctive multiplication cycle ('elementary body' phagocytoses and reorganizes to form large non-infectious 'reticulate body', which divides to form daughter cells which condense into infective elementary bodies); cause, e.g., psittacosis, trachoma and inclusion conjunctivitis.

Spirochaetes

Distinctive, elongate, helically spiral, gram-negative chemoheterotrophs with a flexible cell wall; without external flagella but with between two and several hundred internal flagella ('axial filaments') winding around body in a space within the cell wall, one series of flagella arising from each pole and overlapping in central region, each extending some 67 per cent of cell length; cells single, multiplying by binary fission; cells move in variety of complex locomotory patterns involving bending in loops or coils.

LEPTOSPIRES

Small, obligate aerobes with very few endoflagella; respire a complex range of organic compounds; in aquatic systems or parasitic, causing, e.g., leptospirosis and infectious jaundice.

SPIROCHAETALES

Medium-sized, obligate anaerobes or facultative aerobes fermenting carbohydrates and amino acids; move with characteristic lashings and bendings of cell; free-living in anaerobic or microaerobic sediments, or parasitic, causing, e.g., syphilis and yaws.

CRISTISPIRES

Large (up to 0.5 mm long) obligate commensals with several hundred endoflagella which often form surface crests; many with cross-striations caused by ovoid occlusions; move in smooth forwards/backwards plane; in crystalline style of the mollusc alimentary tract or in guts of wood-eating insects.

Gliding bacteria

Aerobic, gram-negative chemoheterotrophs with flexible cell walls; rod-shaped, occurring singly or in colonies; cells or whole colonies can glide along surfaces although no locomotory organelles occur (small intracellular fibrils may permit the movement); often imbedded in polysaccharide slime; multiply by binary fission.

MYXOBACTERIA

Small rods forming flat, spreading, irregular colonies in thick slime; under certain environmental conditions, vegetative cells aggregate to form macroscopic, often coloured (with carotenoids), fruiting bodies; in fruiting bodies, cells form resting stages (myxospores) which germinate eventually to form motile vegetative rods which aggregate again into colonies; fruiting bodies sessile or on simple or branched stalks; myxospores sometimes forming refractile cysts; many bacteriolytic, killing other microorganisms by production of antibiotics and digesting them extracellularly, others decompose cellulose; mostly in soil or animal and plant debris, a few in freshwaters.

CYTOPHAGAS

Slender, single rods, sometimes occurring in chains; do not form fruiting bodies, may or may not form resting cysts; some facultatively anaerobic, fermenting carbohydrates to organic acids; mostly aerobic decomposers of cellulose or chitin; in soil and aquatic systems, some parasitic in fish.

FILAMENTOUS GLIDERS

Straight or helical filaments, either sessile and attached but releasing motile cells, or whole filament motile; many multiply by random fragmentation of filaments; some oxidize H_2S with deposition of intracellular sulphur granules (these may be partially chemoautotrophic but appear also to require organics such as acetate), others purely chemoheterotrophic; in aquatic systems and in oral cavities of animals.

Endospore-formers

Gram-positive rods forming endospores (endospore is a highly resistant, cryptobiotic, thick-walled resting stage containing a central cell enclosed by a peptidoglycan envelope and an outer spore coat, and capable of surviving for decades); obligate aerobes, obligate anaerobes or facultative anaerobes, respiring or fermenting a wide range of organic compounds including cellulose; multiply by binary fission; mostly motile with peritrichous flagella; mainly in soil, also in aquatic systems, some pathogenic (usually by toxin production) in insects and vertebrates, causing, e.g., anthrax, tetanus and gangrene.

Gram-positive asporogenous rods and cocci

Gram-positive rods or cocci occurring singly, in packets or in chains, which do not produce endospores; aerobic, anaerobic, oxygen-tolerant, or facultatively anaerobic chemoheterotrophs; mostly non-motile; multiply by binary fission.

MICROCOCCI

Aerobic cocci, occurring singly or in clusters (often of 4 cells); with respiratory metabolism utilizing simple organic compounds; some facultatively anaerobic, fermenting sugars; in soil, air and water, some commensal or parasitic on, or in, animals.

LACTOBACTERIA

Rods and cocci with anaerobic fermentative metabolism but capable of tolerating the presence of environmental O_2; do not synthesize haeme proteins; with very limited synthetic abilities; ferment carbohydrates producing lactic acid; occur on and in plants and plant products, and in animals, especially in naso-alimentary systems and the vagina of endothermic vertebrates; some pathogenic, causing, e.g., pneumococcal pneumonia; others important in preparation of dairy products.

ANAEROBIC COCCI

Obligate anaerobes fermenting amino acids and sugars to acetate, formate, etc. and to CO_2 and H_2; abundant in rumen of ruminant mammals and as non-pathogenic symbionts of endothermic vertebrate guts generally.

Actinobacteria

Diverse group of gram-positive chemoheterotrophs varying from single, rod-shaped forms of irregular but often club-shaped outline to fungus-like mycelial colonies; taxonomy of component groups chaotic.

CORYNEBACTERIA

Basically rod-shaped but of variable, irregular form; facultatively anaerobic with both respiratory and fermentative metabolism and complex nutritional requirements; do not form colonies; commensal or parasitic in animals, causing, e.g., diphtheria.

PROACTINOMYCETES

Aerobic respiratory forms producing a rudimentary or transient mycelium which fragments into non-motile rods; some can oxidize petroleum hydrocarbons; colonies often with dry, wrinkled surface; in soil and aquatic systems, and parasitic in animals, causing, e.g., tuberculosis.

ARTHROBACTERS

Aerobic respiratory rods, cocci or myceloid forms with binary and multiple fission and coccoid and rod shapes during different stages of life-cycle; respire a wide range of simple organic compounds; in soil and aquatic systems, commensal and parasitic in plants, and in dairy products.

AIR-TOLERANT ANAEROBES

Single cells of irregular outline, branched cells or rudimentary mycelia; with fermentative metabolism, fermenting lactate to propionate and acetate, or sugars to lactate, formate and succinate; with complex nutritional requirements; on skin and in alimentary tracts of mammals as symbionts or parasites, causing, e.g., acne, and in some dairy products.

DERMATOPHILS

Single cells with two different growth phases determined by nutrition: motile flagellated rods multiplying by budding (the 'R phase') or aggregates of cocci multiplying by fission (the 'C phase'); with aerobic respiratory metabolism; in soil, possibly in brackish waters, and pathogenic in animals.

EUACTINOMYCETES

Colonial aerobic forms with permanent, large (often macroscopic), much branched mycelium lacking cross walls; produce resistant, unicellular actinospores on aerial hyphae arising from vegetative mycelium, on specialized aerial mycelia, or in 'sporangia' borne on an aerial or the vegetative mycelium; some spores with polar flagella, others non-motile; respire a variety of carbohydrates including cellulose and chitin; some produce antibiotics and can hydrolyse peptidoglycans; one form anaerobic, fermenting cellulose in gut of termites; in soil and aquatic systems.

Mycoplasms

Minute (0.3–0.9 μm diameter), non-motile, coccoid, pear-shaped, helical or filamentous chemoheterotrophs primarily and/or secondarily lacking a cell wall and bounded only by a lipid plasma membrane; gram-negative obligate anaerobes or facultative aerobes with fermentative metabolism and complex nutritional requirements including, e.g., cholesterol or other sterols; with very small quantities of genetic material; multiply by binary fission; colonies grown on solid media (e.g., agar) have characteristic 'fried-egg' appearance with dark central and light peripheral area; parasitic in plants and animals.

KINGDOM PROTISTA

MICHAEL A. SLEIGH
JOHN D. DODGE
AND
DAVID J. PATTERSON

This kingdom comprises eukaryote microorganisms and their close descendants. All members of the Animal and Plant Kingdoms, and probably also the fungi, are assumed to have evolved from protistan ancestors, and it is simplest to regard this Kingdom as comprising all the eukaryotes that are not members of the Animal, Fungal or Plant Kingdoms, as defined later. It thus includes organisms traditionally classed as algae and protozoa, as well as fungi with flagellated spores, but not microscopic metazoans like mesozoans or microscopic fungi like yeasts, whose organization indicates their membership of other Kingdoms. In common with other eukaryotes, they have a nuclear envelope, cytoplasmic microtubules which are concerned in nuclear division, and usually a flagellum, whose '9+2' axis is formed of microtubules; mitochondria are usually present and plastids are often possessed.

Among microscopic eukaryotes are autotrophic or auxotrophic, oxygenically photosynthetic forms whose plant-like nature was recognized by early microscopists; non-photosynthetic, heterotrophic forms also abound and early observers regarded these as animals or sometimes fungi. An apparently reasonable extension to the simpler eukaryotes of the obvious separation between Animal, Fungal and Plant Kingdoms gives rise, however, to unresolvable conflicts in the classification of many organisms. For example, amongst euglenids less than one-third of the genera are photosynthetic, and hence plants, whilst the remaining genera are heterotrophs, and hence animals (or fungi!). Multicellular animals, non-flagellate fungi and the embryophyte land plants represent major terminal branches (or groups of branches) of the eukaryote section of the evolutionary tree, recognizable as more or less distinct Kingdoms of organisms. Over recent decades it has become increasingly clear that there also exist many types of eukaryotes, representing separate branches from the trunk of the evolutionary tree, that are not assignable to any one of these three Kingdoms. These other branches represent different degrees of specialization, and some of them may even be thought sufficiently isolated for them to be recognized as Kingdoms in their own right. However, since all organisms are part of the same evolutionary tree, each branch being connected to all others, the artificial division of the tree into Kingdoms, which must be arbitrary, becomes especially problematical when one considers smaller twigs from the main trunk. It is, therefore, accepted that, for the time being, we should regard all of these smaller branches representing simpler organisms as belonging to the one Kingdom, that is, Kingdom Protista.

The evolution of protists has taken place at the cellular level and differences in the nature and arrangement of the basic cellular components and their products are the features that differentiate the different groups of protists, and allow us to characterize each group by its ultrastructural inventory. The patterns of organization of the different groups suggest relationships, e.g., the similarity of euglenids and kinetoplastids, but yet emphasize the separate identity of such groups within the same Kingdom as, perhaps, neighbouring branches of the evolutionary tree. We have, therefore, regarded each of the major distinct patterns of organization found among protists as constituting a phylum, and have named more phyla than in many classification schemes.

Nearly all of the major groups of protists bear '9 + 2' flagella, as do most phyla of both the Animal and Plant Kingdoms, and it is widely assumed that the ancestral eukaryotes were flagellated; the presence of flagella is, therefore, not regarded as a feature that separates 'flagellates' from other organisms, although the number and arrangement of flagella and associated structures provide distinctive characteristics of many major taxa. Similarly, many of the major groups of protists possess chlorophyll-containing plastids, but these are of many different types, in both their structure and the pigments they contain, and they are thought to have been independently acquired as symbionts by various heterotrophic flagellates; hence there is no reason to regard photosynthetic groups as more closely related to each other than to other protist groups. Furthermore, most naked cells show amoeboid tendencies, so the fact that cells produce pseudopodia does not mean that they are closely related and, therefore, that all 'amoebae' have a common origin.

Important characters to consider in distinguishing between protistan phyla therefore will include features of flagella, plastids and pseudopodia, and also variants of most other cell components. Cells of many phyla produce a protective coat outside the cell membrane, whose nature and composition is distinctive, and cells of other phyla develop various characteristic arrays of structures on the inner side of the cell membrane, as a pellicle. The structure of the nucleus and its chromosomes, the behaviour of the nuclear membrane, chromosomes and mitotic spindle at karyokinesis and the form of structures at the spindle poles are surely fundamental variables, and the occurrence and form of meiotic divisions and their place in the life-cycle are, in addition, often distinctive. Different modes of nutrition are associated with different cell organelles and with the accumulation of distinctive storage products. Metabolic differences may also determine the presence or absence of mitochondria, and there are also differences in the form of the internal cristae of the mitochondria that characterize various groups. Many other distinctive features of skeletal structures, golgi bodies or other cell components characterize specific taxa. Within most phyla there are colonial or other multicellular members, and the form of multicellular aggregates and the nature of the relationships between component cells can be distinctive.

Although it is normally possible to identify an individual protist with the light microscope, time-consuming electron microscopy is required to elucidate many of the diagnostic features that confirm its phyletic relationships. It is not too surprising, therefore, that we are only slowly gaining information about enough members of each phylum, class or order to know which features vary and which are consistent within that taxon. Knowledge is continually accumulating, and conclusions reached here on the basis of examinations of a few species are likely to be refined later as more representatives are examined. There are some small groups, possibly consisting of only one or two species, which appear, on present evidence, to be very different from any other organisms, and which we have found it difficult to allocate to any larger phylum; in some cases these organisms have themselves been well-characterized, but their possible relatives, assigned to other groups, have not been studied; in other cases the isolated organisms themselves are not known in sufficient detail to indicate their relationships. Sometimes we have felt too ignorant to suggest any relationship for such groups or species (they are listed on pp. 87–88), and on other occasions we have used weak evidence to support tentative conclusions about relationships, and have indicated this.

The grouping of recognizable species into genera and families usually gains a measure of acceptance by other workers, but criteria for grouping families into orders, classes and phyla are much more debatable. Among the Protista the classification of the majority of organisms up to the level of orders will be recognized, if not entirely approved, by the majority of workers in the field; it is these orders that form the lowest category considered here. In bringing together these orders into a uniform scheme we have met with several difficulties. Firstly, our conclusions about the classes and phyla to which orders of protist belong do not always coincide with those for which there is published precedent. It is our intention to provide a practical and convenient scheme which recognizes the similarities and differences between organisms that are now known to exist. Previous schemes of classification do not, in our opinion, take account of recent advances in understanding, particularly among flagellated and amoeboid groups, although the recent recognition of some new phyla in other areas has somewhat clarified this. Some new phyla and classes have been proposed where this is felt to be justified: their names will be recognized from past usage, and we expect further evidence will lead to their proper establishment in the near future. The other major problem concerns conflicts between botanical and zoological practices in nomenclature, both in terms of the names of higher categories of classification and in the standard endings for these categories. We feel that we should be as consistent as possible within the Kingdom Protista, and, in general, zoological practice is followed, except where established botanical names are more familiar.

Nearly 120 000 species of protist have been described, rather less than one-half of these being fossil forms. It is certain that many more species, particularly from marine and endobiotic habitats, remain to be described. By virtue of their naked surfaces, active protists are limited to damp environments, being found in water or damp soil or as endobionts.

The 27 phyla of protists listed here can be assumed to have evolved a very long time ago, in the early radiation of eukaryotes. Since the organisms have continued to evolve, we may find it difficult to prove relationships between phyla; perhaps only molecular methods, like ribosomal RNA sequencing, will help to confirm or deny any suggested relationships. However, we have attempted to list the phyla in a sequence which places close together phyla that share common characters that we feel may reflect phylogenetic affinity.

Phylum Parabasalia

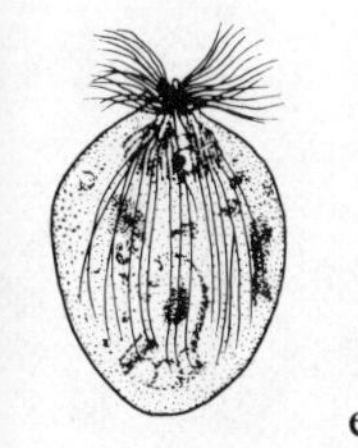

These are flagellates, usually with several or many flagella borne on the anterior part of the body, in which at least some of their basal bodies are associated with parabasal structures composed of striated filamentous fibres with closely applied golgi complexes. Cytoplasmic microtubular elements also arise from basal bodies. Neither plastids nor mitochondria are present, and there is neither a cell wall nor a specialized pellicle. The nuclear membrane remains intact (closed) during mitosis, the spindle being extranuclear and extending across one side of the nucleus, with chromosomes connected to microtubules at kinetochores situated in the nuclear envelope; the spindle poles are associated with 'atractophores' (fibrous extensions from certain basal bodies). The chromosomes remain condensed during interphase. The Parabasalia are all or nearly all endobionts and are mostly phagotrophic. A single class contains 2 orders and perhaps 2000 species.

ORDER TRICHOMONADIDA

One to many karyomastigonts (groups of flagella associated with a nucleus); karyomastigonts usually have 4–6 flagella, one turned back as a recurrent flagellum and often attached to body surface to form an undulating membrane, other flagella free; flagella occasionally reduced in number or missing; flagellar bases of each karyomastigont associated with 2 microtubular organelles, a short pelta and a longer, non-contractile, axostyle (which often ensheathes the nucleus), as well as with the filamentous parabasal fibre and other structures; with hydrogenosomes; mostly endobionts in animals, including mammals. 4 families.

ORDER HYPERMASTIGIDA

Many flagella present, but single nucleus; flagella not arranged as distinct karyomastigont groups, but 1–2 groups of 'privileged' basal bodies (lacking flagella) may be specifically associated with the multiple parabasal fibres and the 'atractophore'; locomotor flagella occur in longitudinal or spiral rows, in a circle or in 'plates'; sexual reproduction reported in several genera, some haploid, some diploid; all endobionts in insects, mostly xylophagous. 11 families.

Phylum Dinophyta

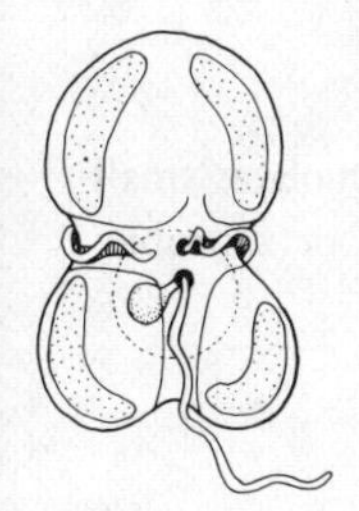

These are mainly unicellular organisms with two dissimilar flagella, one being helical or undulating and with a paraxial rod, the other of the whiplash type. The cell covering is a pellicle formed from a complex theca, which may be composed of distinct internal cellulosic plates that provide important taxonomic criteria, underlain by microtubules. Chloroplasts, when present, are brownish or golden, with chlorophylls *a* and c_2, β-carotene and various xanthophylls; thylakoids normally arranged in threes. Food reserves are starch and oil. Mitochondria (with tubular cristae) and golgi bodies are present. Nucleus conspicuous, containing condensed chromosomes. During mitosis the nuclear membrane remains closed and the spindle is extranuclear; usually without centrioles. Most are haploid organisms in which sex involves flagellate isogametes. Trichocysts present. Some with an eyespot, whose position varies with species. Many species phagotrophic or saprobic. Aquatic organisms, both marine and freshwater, also found as symbionts (e.g., in corals) and parasites. 'Dinoflagellates'. About 2000 living species and some 2000 fossil species in a single class, the Dinophyceae.

ORDER PROROCENTRALES

Flagella inserted apically; cell mainly covered by 2 large thecal plates; golden-brown chloroplasts present; mainly planktonic but some interstitial amongst sand. 1 family.

ORDER DINOPHYSIALES

Flagella inserted at one side towards anterior end of cell; distinct transverse groove (girdle) present, sometimes bounded by broad flaps (lists); hypotheca mainly covered by 2 large thecal plates; chloroplasts usually present. 2 families.

ORDER GYMNODINIALES

Cell appears naked, readily bursts; distinct girdle and sulcus (longitudinal groove) present, but position and orientation variable; some contain chloroplasts (of various colours), others phagotrophic or saprobic; some produce toxins. 5 families.

ORDER NOCTILUCALES

'Naked' cells, usually very large (> 1 mm) and highly vacuolate; small tentacle present; lack chloroplasts; nutrition phagotrophic or saprobic; may be bioluminescent. 1 family.

ORDER PERIDINIALES

Cell covered with theca of cellulosic plates with distinctive shapes and ornamentation; sulcus and girdle usually conspicuous; many contain yellow-brown chloroplasts; some form resistant cysts; some produce toxins. 18 families.

ORDER BLASTODINIALES

Main phase parasitic (usually ectoparasitic) on metazoans or copepod eggs, although chloroplasts may be present; motile infective stage gymnodinioid or gonyaulacoid; cyst stages in life-cycle of some. 7 families.

ORDER SYNDINIALES

Endoparasitic, host usually another dinoflagellate, protozoan or metazoan; large nucleus with very few chromosomes and distinctive nuclear division; chloroplasts absent. 4 families.

ORDER PHYTODINALES

Little-known; main phase coccoid or attached to substratum; motile zoospores gymnodinioid; chloroplasts present. 2 families.

Phylum Kinetoplasta

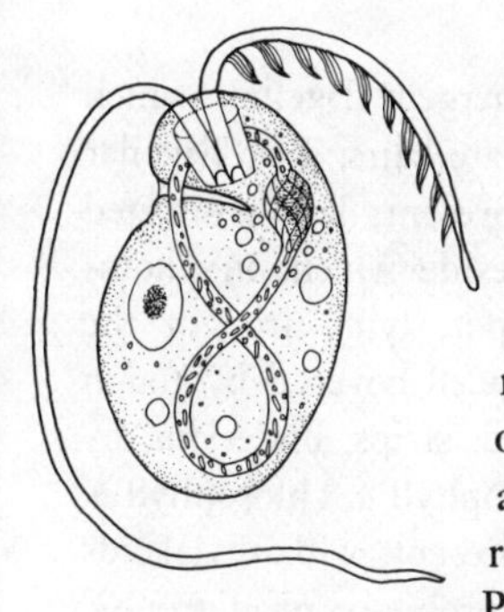

These flagellates have one or two flagella that emerge from a depression or pocket in the naked body surface. Typically there are two parallel basal bodies linked by filaments; the only associated rootlets are a quartet of microtubules that run alongside the flagellar pocket and join an array of other microtubules in the pellicle. Alongside the axoneme within the flagellar shaft is a paraxial rod (as in the Dinophyta and Euglenophyta). Plastids are not present. The single mitochondrion often extends through much of the cell as a hoop or network and typically contains a large concentration of fibrillar DNA (the kinetoplast) located close to the flagellar basal bodies; the cristae are plate-like. Typically with a cytostome supported by microtubules, but not normally forming pseudopodia. The nucleus is vesicular and contains a prominent endosome; mitosis takes place within a closed nuclear envelope, with an internal spindle, and without obvious polar structures, only small nucleating centres on the inner nuclear membrane. Sexual reproduction has not been confirmed. Golgi bodies are usually found near the flagellar bases, but are not connected to them or to associated fibres. There is often a contractile vacuole which empties into the flagellar pocket. A single class with 2 orders, and about 600 species.

ORDER BODONIDA

Two heterodynamic flagella usually present, typically one trailing and one held forward or laterally; flagella emerge from shallow depression; one or both flagella may bear single row of fine hairs; with well-formed cytostome; kinetoplast often large and single, sometimes dispersed in bodies throughout part or all of mitochondrion; free-living and parasitic. 3 families.

ORDER TRYPANOSOMATIDA

One flagellum, which emerges from deep flagellar pocket, sometimes attached to pellicle as an undulating membrane, sometimes free; second, barren, basal body usually present; kinetoplast small and compact; parasitic in animals, including man, or occasionally in plants. 1 family.

Phylum Euglenophyta

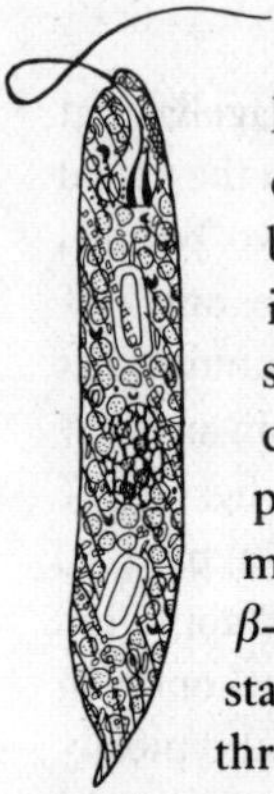

Motile unicells, often with only one emergent flagellum which contains a paraxial rod and bears delicate hairs. Two flagellar bases, with parallel alignment, usually present; flagella inserted into anterior depression (reservoir) beside which, in photosynthetic forms, is situated an eyespot, lying free in the cytoplasm, and a contractile vacuole. Cell covered by tough pellicle made up of spirally arranged strips underlain by microtubules. Green chloroplasts (chlorophyll *a*, chlorophyll *b*, *β*-carotene and xanthophylls) may be present, with thylakoids stacked in threes, or nutrition may be saprobic or phagotrophic through a cytostome beside the reservoir. Large refractile paramylon grains form food reserve. Mitochondria (with discoidal cristae) and golgi present. Conspicuous nucleus with permanently condensed chromosomes. Nuclear membrane remains closed during mitosis and its spindle is entirely endonuclear, without centrioles. Mainly freshwater organisms, particularly common in sites with high organic nutrients. About 1000 species in a single class, the Euglenophyceae.

ORDER EUTREPTIALES

With 2 emergent flagella, one directed forward, the other backward, both beat actively; some possess chloroplasts, others do not but never phagotrophic; non-rigid body. 1 family.

ORDER EUGLENALES

Only 1 of the 2 flagella emerges from anterior canal; most possess chloroplasts, but some saprobic; generally, cell covering very flexible. 1 family.

ORDER RHABDOMONADALES

Only 1 flagellum emergent; without chloroplasts, nutrition saprobic; cell covering rigid. 1 family.

ORDER HETERONEMATALES

With 1–2 emergent flagella, one always stretched out in anterior direction with only tip moving when cell swims; without chloroplasts, nutrition phagotrophic; ingestion apparatus present. 1 family.

ORDER SPHENOMONADALES

With 1–2 emergent flagella, one directed forwards, held rather rigidly; no chloroplasts present; nutrition saprobic or phagotrophic; no special ingestion organelles. 1 family.

ORDER EUGLENOMORPHALES

With 3 or more emergent flagella, all of same length; chloroplasts generally not present; in gut of tadpoles. 1 family.

Phylum Opalinata

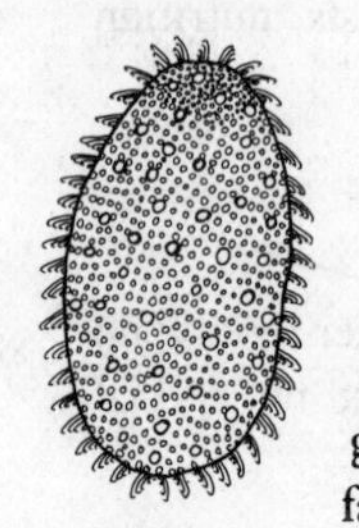

Exclusively endosymbiotic protists without chloroplasts, found in the digestive tracts of amphibians and fish; move by action of undulating flagella disposed in rows (kineties) extending over the cell body. Body division may occur along or across kineties, additional kineties may be added during growth at a differentiated region of one anterior margin, the falx. Most species are flat, with two to hundreds of similar nuclei; the mitochondria have tubular cristae; and the pellicle is ribbed and supported between kineties by microtubules. Diploid with syngamy of multiflagellate gametes; cysts formed. A single class, order and family containing about 400 species.

Phylum Stephanopogonomorpha

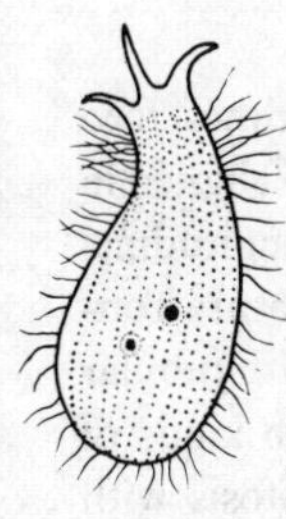

Organisms found in marine interstitial habitats with a polar mouth surrounded by lappets, and moved by the action of flagella lying in rows (kineties) over the body surface. With homokaryotic nuclei (cf. ciliates with which they were for a long time classified); mitochondria with flattened cristae; nucleus with internal division spindle. Sexual reproduction not known. Only a few species in 1 genus.

Phylum Ciliophora

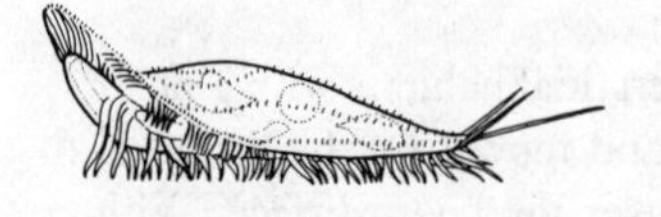

Protists which move and sometimes feed using cilia; locomotor cilia usually aligned in rows (kineties) which define the longitudinal axis of the body and whose infraciliature contributes to a complex pellicle; without plastids; mitochondria with tubular cristae; golgi system present but with few adpressed sacs; two kinds of nuclei—one a diploid micronucleus capable of meiosis or mitosis with spindle located inside the intact nuclear envelope, and without centrioles, and the other a selectively polyploid macronucleus unable to undergo meiosis or true mitosis, but in most groups dividing amitotically; sexual reproduction typically involves exchange of nuclei at conjugation; cysts in some species; some species with endosymbionts; growth does not involve increase in numbers of kineties, division occurs across the kineties; typically with a single site, the cytostome, for food ingestion; cytostome may lie at cell surface or in a depression with which are associated compound ciliary organelles for feeding; nemadesmal rods (compound microtubular structures) and/or extrusomes often lie internal to the cytostome. 'Ciliates'. Classification unstable, 7500 species in 3 classes.

Class Kinetofragminophorea

Oral ciliature absent or, if present, only slightly distinct from somatic ciliature; cytostome terminal, lateral, subterminal or ventral, rarely in a depression; generally not filter-feeders, but scavengers, predators, algivores etc.

Subclass Gymnostomatia

Cytostome typically located at or near anterior pole of cell and not in depression; without well-developed oral ciliature; mainly predatory with extrusomes used in prey capture. 3 orders.

ORDER KARYORELICTIDA

Macronuclei not capable of division (even by amitosis); mouth typically concave, adjacent to anterior pole; mostly interstitial organisms; postciliary rootlets from bases of somatic cilia very strongly developed (as in heterotrichs, see below). 3 families.

ORDER PROSTOMATIDA

Mouth at front pole, feeding on detritus or predatory, consequently with well-developed nemadesmata and/or killing/holding extrusomes. 11 families.

ORDER PLEUROSTOMATIDA

Flattened; carnivorous; cytostome slit-like, extending from front pole along one edge of body; extrusomes associated with mouth. 1 family.

Subclass Vestibuliferia

Cytostome located in depression; many are suspension feeders, using tightly-packed cilia derived from anterior ends of somatic kineties to create feeding currents; many endobiotic. 3 orders.

ORDER TRICHOSTOMATIDA

Body typically evenly ciliated, mouth cilia incorporating only cilia from kineties and a few super-numerary fragments of kineties; includes endobionts of vertebrates (incl. man). 11 families.

ORDER ENTODINIOMORPHIDA

Body typically lacking regular kineties; somatic and buccal cilia in form of tufts of cilia (syncilia); pellicle rigid and often drawn out into points or folds; commensals in herbivorous mammals; ingest bacteria, plant tissue or other protists. 7 families.

ORDER COLPODIDA

Cilia in pairs, kineties spirally arranged; vestibular cilia densely packed into well-organized arrays; feeding on bacteria or on other protists; many common in soils. 3 families.

Subclass Hypostomatia

Mouth typically located away from anterior pole, often on flattened ventral surface; mouth often with well-developed nemadesmata; morphogenesis, particularly stomatogenesis, preceding division is complex. 7 orders.

ORDER SYNHYMENIDA

Free-living, mainly from freshwater habitats; mouth with strongly developed nemadesmata; ciliation even; body typically cylindrical; continuous row (frange) of cilia leading away from mouth towards anterior pole of cell. 3 families.

ORDER NASSULIDA

Free-living, mainly in freshwaters; mouth with strongly developed nemadesmata; row of cilia leading away from mouth (frange) broken into units, sometimes reduced to a few membranelle-like structures; often with suture leading to mouth from anterior end of cell; often algivorous. 5 families.

ORDER CYRTOPHORIDA

Somatic ciliature typically restricted to flattened ventral surface, sometimes cilia in 2 distinct fields; nemadesmata often well-developed, with teeth-like capitula at their distal ends; often thigmotactic and with posterior adhesive organelle; feeding typically on bacteria adhering to surfaces. 8 families.

ORDER CHONOTRICHIDA

Typically vase-like and sedentary, ectocommensal on crustaceans; mostly marine; without evident somatic cilia but with cilia at anterior end used for feeding surrounded by lips or spiral body fold; superficially unlike preceding taxon, but similarities revealed through patterns of morphogenesis and disposition of cilia on motile larval chonotrichs. 11 families.

ORDER RHYNCHODIDA

Adult form lacking cilia (usually) but with rostrum, used as sucking device; mainly described from gills of marine molluscs; motile larval form. 2 families.

ORDER APOSTOMATIDA

Mainly associated with marine crustaceans, particularly active during moulting; complex life-cycle; glandular rosette associated with mouth at some stages of life-cycle, but mouth absent from some; somatic cilia in widely spaced spirals; often thigmotactic. 5 families.

ORDER SUCTORIDA

Typically sedentary, lacking somatic cilia but with 1 to many tentacles (mouths) through which food is sucked; mainly predatory, often feeding on other ciliates; motile ciliated larva produced by budding. 19 families.

Class Oligohymenophorea

Typically with cytostome at base of depression in cell surface (buccal cavity) near front of cell or located ventrally; typically with small number of compound ciliary organelles used in feeding and distinct from somatic cilia; mainly suspension feeding. 2 subclasses.

Subclass Hymenostomatia

Even body ciliature (usually).

ORDER HYMENOSTOMATIDA

Oral ciliature relatively simple, typically with 3 blocks of tightly packed cilia to left of mouth and line of cilia to right; includes many familiar ciliates such as *Tetrahymena* and *Paramecium*. 14 families.

ORDER SCUTICOCILIATIDA

Resembling previous order in many respects, but line of cilia to right of mouth strongly developed and peculiar 'scuticus' stage occurs during stomatogenesis. 23 families.

ORDER ASTOMATIDA

Mouthless endoparasites of invertebrates, mainly oligochaete annelids; often occurring as chains of incompletely separated organisms. 10 families.

Subclass Peritrichia

Oral cilia form wreath around oral end of cell, somatic cilia reduced to basal wreath or absent; many sessile, some of which are colonial; some symbiotic, living on surfaces of soft-bodied marine and freshwater invertebrates. A single order (Peritrichida), 17 families.

Class Polyhymenophorea

Buccal cilia in form of band of membranelles leading from front pole of cell to cytostome, typically located on ventral surface of body; some with line of cilia to right of mouth; typically filter feeders; somatic ciliature sometimes greatly reduced. 4 orders.

ORDER HETEROTRICHIDA

Even ciliature of body, contrasting in conspicuity with membranelles; like karyorelictidans (see p. 40), have strongly developed postciliary rootlets associated with kinetosomes; often very large, some strongly pigmented. 19 families.

ORDER ODONTOSTOMATIDA

Laterally flattened with poorly-developed adoral zone of membranelles (less than 10 membranelles); found mainly in association with putrefying organic matter in freshwater habitats. 3 families.

ORDER OLIGOTRICHIDA

Mainly marine pelagic ciliates with well-developed zone of membranelles used for locomotion as well as feeding; includes the widespread predominantly marine 'tintinnids'. 15 families.

ORDER HYPOTRICHIDA

Generally flattened, with locomotor cilia usually restricted to ventral side and present in form of packed aggregates, cirri, which may be present in rows resembling kineties, or may be more sparsely distributed; cirri act as 'legs' to push cell over substratum. 11 families.

Phylum Apicomplexa

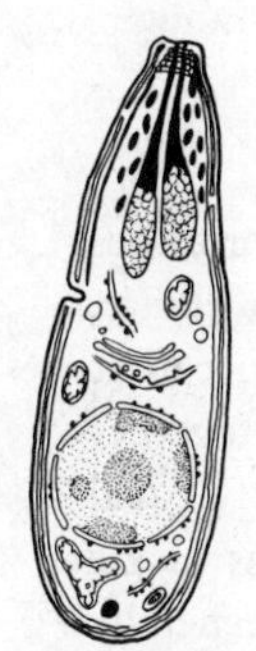

Parasitic protists without plastids, but usually with mitochondria (having vesicular cristae) and golgi bodies. Life-cycles with several stages, at least one of which normally possesses an apical complex of organelles. This consists of a conoid (a spiral band of microtubules shaped as a truncated cone) surrounding the necks of flask-shaped rhoptries and/or more slender micronemes (both apparently secretory bodies), and surrounded by one or two polar rings which are associated with the pellicle (typically of three membrane layers) and pellicular microtubules. One or more stages also possess surface specializations called micropores, which may be sites of pinocytosis. Apicomplexans are haploid with sexual reproduction by syngamy. Division is characteristically by schizogony, many nuclear divisions producing a multinucleate cell before the cytoplasm segments; nuclear divisions are closed with an intranuclear spindle, or with polar fenestrae, but centrioles are not usually present. Flagella are absent except for those on microgametes of some species and on the 'zoospores' of one other species. About 5000 species, mostly in 1 class.

Class Perkinsea

Conoid forms incomplete cone, open at one side; no sexual reproduction; free-living migratory 'zoospore' stage (possibly homologous with sporozoite stage of the Sporozoea) with anterior flagellum with row of fine hairs, naked posterior flagellum, and prominent anterior vacuole; homoxenous (only 1 host in life-cycle). 1 species only.

Class Sporozoea

Conoid (when present) forms complete, but truncated, cone; sexual and asexual reproduction in life-cycle; each zygote normally forms an oocyst wall within which it undergoes meiosis, sometimes followed by mitosis, in a process of sporogony producing mobile vermiform sporozoites (infective stage); mitotic divisions also occur during merogony of feeding stages (trophozoites) and during gametogony; microgametes of some flagellated;

locomotion of other gametes and any other motile stages by gliding or body flexion; some form pseudopodia, but use them only in phagocytosis; homoxenous or heteroxenous (more than 1 host in life-cycle).

Subclass Gregarinia

Life-cycle typically includes only gametogony and sporogony; sporozoites penetrate host cells, but trophozoites grow large and become extracellular, usually maintaining connection with host cells by thorn-like mucron or expanded anterior epimerite; pairs of mature cells (gamonts) come together (syzygy) prior to undergoing gametogony, which produces nearly equal numbers of similar gametes in each gamont (male gamete has single emergent flagellum in some); syngamy within gametocyst; generally homoxenous in gut or body cavity of invertebrates or lower chordates.

ORDER ARCHIGREGARINIDA

Typically with merogony, gametogony and sporogony in apparently primitive life-cycle; in annelids, sipunculans, hemichordates and ascidians. 2 families.

ORDER EUGREGARINIDA

With gametogony and sporogony, but lacking merogony; gamont usually divided into 2 compartments by septum; usually in annelids or arthropods. 25 families.

ORDER NEOGREGARINIDA

In gut, malpighian tubules, fat body or haemocoel of insects, where they undergo merogony (apparently secondarily reacquired) as well as gametogony and sporogony. 5 families.

Subclass Coccidia

Life-cycle typically with merogony, gametogony and sporogony; mature gamonts small and usually intracellular without mucron or epimerite; usually without syzygy; commonly in vertebrates, including domestic animals and man.

ORDER AGAMOCOCCIDIIDA

With neither merogony nor gametogony; in annelids. 1 family

ORDER COELOTROPHIDA

Without merogony and developing extracellularly; in annelids. 5 families.

ORDER EUCOCCIDIIDA

With merogony; homoxenous in invertebrates or vertebrates or heteroxenous in 2 invertebrates or in 1 invertebrate and 1 vertebrate; syzygy of macrogamete with microgamont occurs in 1 suborder; includes parasites causing malaria and coccidiosis. 8 families.

Subclass Piroplasmia

Minute rounded or pyriform parasites found within erythrocytes, or other circulating or endothelial cells of vertebrates, where they reproduce by merogony; known vectors are ticks, in which they undergo sporogony; without oocysts or spores; apical complex with polar ring and rhoptries but without conoid and usually without associated pellicular microtubules; flagella absent, trophozoite stages separated from erythrocyte cytoplasm only by single membrane (usually at least 2 membranes in other sporozoans); sexual reproduction probably occurs in the tick. 1 family.

Phylum Metamonada

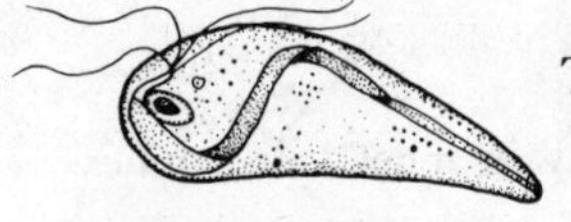

These flagellates have one or several karyomastigonts, each of which usually has one or two pairs of flagella, the two basal bodies of each pair arranged orthogonally, and with at least one flagellum of each karyomastigont being typically turned back as a recurrent flagellum. Microtubular rootlets from the basal bodies form bundles passing into the cytoplasm, and microtubular and/or striated root fibres connect the basal bodies to the nuclei. The mitotic spindle is normally intranuclear, within a persistent nuclear envelope, the flagellar basal bodies lying outside the nucleus near to small plaques in the inner membrane at the spindle poles (polar fenestrae in some species). Plastids, mitochondria and golgi apparatus are all absent; the cell surface is naked, but pellicular microtubules may be present. Mostly endobiotic and nutrition usually phagotrophic; cysts formed. 2 classes containing a few hundred species.

Class Anaxostylea

Without microtubular axostyle; typically with ventral 'cytostome area' bordered by fibres (principally microtubular) originating from flagellar bases, and crossed by a recurrent flagellum. 2 orders.

ORDER RETORTAMONADIDA

Parasitic; with single karyomastigont and 2–4 flagella; pellicle with many single microtubules. 2 families.

ORDER DIPLOMONADIDA

Typically with 2 karyomastigonts, each with 4 flagella, the 2 karyomastigonts arranged in bilateral or 2-fold rotational symmetry, and sometimes 2 cytostomes; single karyomastigonts with 1–4 flagella occur in some; parasitic or occasionally free-living. 2 families.

Class Axostylea

With 1 to many karyomastigonts; the 2 pairs of basal bodies of each karyomastigont associated with a paracrystalline preaxostyle connected to anterior end of a thick ribbon of microtubules—the axostyle; this type of axostyle is often motile, showing propagated undulations; recurrent flagella may adhere to body surface for greater or lesser distance; sexual reproduction reported in several species, some haploid, some diploid; all endobionts in insects. 1 order (Oxymonadida), 4 families.

Phylum Sarcodina

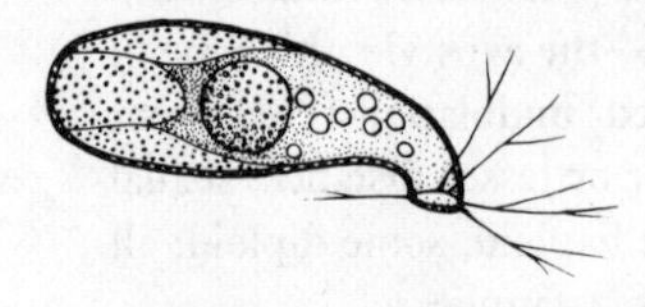

The amoeboid protists, which are characterized by their ability actively to produce transient extensions of the cell surface (pseudopodia) which may be used for locomotion or feeding. Mostly with mitochondria, golgi complexes and microtubules, but many also with contractile vacuoles and some with extrusomes and/or endosymbiotic algae. A diversity of mitochondrial cristal form and of nuclear division profiles are encountered. Some species produce a flagellated form either in response to environmental factors or as part of the life-cycle. Many have organic or inorganic internal skeletons or external tests. Many produce spores or cysts. No chloroplasts (except in their symbiotic algae). Probably polyphyletic; about 48 000 species of which about two-thirds are fossil and most are free-living. 2 subphyla.

Subphylum Rhizopoda

Amoebae with relatively active pseudopodia not supported internally by stable arrays of geometrically packed microtubules; pseudopodia mostly broad (lobose), thin (filose) or forming a branching network (reticulopodia); mitochondria usually with tubular or vesicular cristae; some species naked, some with tests or skeletons, and some able to produce flagella or cysts. 41 000 species in 9 classes.

Class Lobosea

Typically with broad (lobose) pseudopodia; mainly free-living. 4 orders.

ORDER GYMNAMOEBIDA

Lobose; without any evident (visible with light-microscope) cell coat or lorica; with 1 pseudopodium (monopodial) or many (polypodial); mostly from soil or freshwater habitats. 9 families.

ORDER SCHIZOPYRENIDA

Lobose; without evident surface coat or lorica; able to produce flagella in response to environmental stress; mostly described from soil; cysts produced. 1 family.

ORDER ARCELLINIDA

Lobose; with test of organic matter or incorporating secreted or agglutinated inorganic particles; test with aperture from which pseudopodia emerge; in soils, mosses and interstitial habitats, in addition to more conventional freshwater and marine habitats; some with symbiotic algae. Species and genera distinguished by form and texture of test. 9 families.

ORDER TRICHOSIDA

Marine; with short conical pseudopodia; probably with polymorphic life-cycle; one stage bearing calcareous spicules on cell surface. 1 family.

Class Pelobiontea

Probably only the one species, *Pelomyxa palustris*, which is monopodial, algivorous and found in microaerophilic habitats. Cytologically simple, without mitochondria but with large numbers of symbiotic bacteria; no flagella but non-motile cilia seen at the ultrastructural level; body typically filled with sand grains; when moving, cytoplasm exhibits a bidirectional fountain-flow. Held by some to be the most primitive eukaryote.

Class Acarpomyxea

Soil and marine amoeboid organisms forming small extended plasmodia having branching and anastomosing channels (pseudopodia), but not forming fruiting bodies; no flagellated stages; some form cysts. 2 orders.

ORDER LEPTOMYXIDA

Fan-shaped, using anastomosing pseudopodial system to capture bacteria and other protists as food; with 1 to hundreds of nuclei. Resemble superficially some slime moulds and regarded by some as lying evolutionarily between the lobose amoebae and the Eumycetozoa. 2 families.

ORDER STEREOMYXIDA

Pseudopodia more branching than anastomosing, cytologically contains a microtubule-organizing bar; mainly described from marine habitats. 1 family.

Class Acrasea

Uninucleate amoebae with lobose pseudopodia which may aggregate and develop a fruiting body extending up from the substratum and within which many individual amoebae may encyst. One species with a flagellated stage. A superficial similarity to fungal fruiting bodies led them, and some other amoeboid organisms, to be regarded at one time as a simple type of fungus, hence the name 'slime mould.' 1 order (Acrasida), 3 families.

Class Eumycetozoa

Slime moulds able to produce uninucleated spores at end of secreted stalk; trophic organisms, initially uninucleate amoebae, with filose subpseudopodia extending from broad pseudopodial front; may grow to a mass of cytoplasm containing, in some cases, millions of nuclei and in which movement often accompanied by ebb and flow of cytoplasm; some members capable of producing flagellated stage; mitochondrial cristae tubular, resembling those of lobose amoebae. 3 subclasses, 550 species.

Subclass Protosteliia

Trophic organisms either uninucleate amoebae or small plasmodia in which ebb and flow motion of cytoplasm not marked; fruiting bodies with few (sometimes 1) spores; some produce flagellated stage and sexuality known in 1 case. 1 order (Protosteliida), 3 families.

Subclass Dictyosteliia

Trophic amoebae aggregate to form pseudoplasmodia in which integrity of each cell maintained (cf. Myxogastria); fruiting body composed of stalk cells which secrete cellulose and cells which encyst to form spores; no flagellated stages known; genetic recombination occurs but mode of sexuality uncertain. 1 order (Dictyosteliida), 2 families.

Subclass Myxogastria

Acellular or 'true' slime moulds; spores release individual amoebae which ultimately transform into multinucleated, typically fan-shaped plasmodia, exhibiting ebb and flow motion of cytoplasm; contain in some cases an estimated 10^8 nuclei; organisms with 2 naked flagella produced at times; meiosis and cell fusion known. 5 orders.

ORDER ECHINOSTELIIDA

Typically with minute sporangia less than 1.5 mm in height; plasmodia small with relatively few nuclei. 2 families.

ORDER LICEIDA

Spore-bearing body light coloured and without a capillitium (meshwork of threads holding spores together); a pseudocapillitium, possibly remains of spore wall, may be present. 3 families.

ORDER TRICHIIDA

Spore mass with capillitium, no calcium present. 2 families.

ORDER STEMONITIDA

Spore mass dark, perhaps ferruginous in some; lime deposits absent from spore mass but may be present at base of supporting stalk. 2 families.

ORDER PHYSARIDA

Spores may be very dark; spore-bearing body incorporates calcareous deposits; plasmodia may measure several cm in length. 2 families.

Class Granuloreticulosea

Pseudopodia have granular appearance and form anastomosing networks; with microtubules (not in geometric arrays) and extrusomes internally, but cytological data sparse. Most members are foraminiferans; affinities of non-foraminiferal amoebae included here are uncertain. 3 orders.

ORDER ATHALAMIDA

Amoebae without test but with extending networks of pseudopodia; freshwater. 1 family.

ORDER MONOTHALAMIDA

With single-chambered organic test; 1 aperture from which pseudopodia extend; alternation of generations not known. 1 family.

ORDER FORAMINIFERIDA

With multichambered test, chambers often spirally arranged, most pseudopodia extending from terminal aperture of largest chamber; test organic (often brown in colour) or may incorporate agglutinated particles or may be calcareous, hyaline or opaque; exclusively marine, in variety of benthic and planktonic habitats; often with large populations of symbiotic algae; some exhibit nuclear dimorphism, some alternation of sexual and asexual generations; 37 500 species of which many known only as fossils. 'Foraminiferans' or 'forams'. About 115 families.

Class Xenophyophorea

Little-known group of marine amoebae with uncertain affinities; multinucleated plasmodial organisms; construct agglutinated tests; body may contain barite ($BaSO_4$) crystals. 2 orders.

ORDER STANNOMINIDA

Test contains linellae, stiffening elements, which may be only a few μm wide but several mm long; body flexible. 1 family.

ORDER PSAMMINIDA

Test without linellae; body stiff. 4 families.

Class Plasmodiophorea

Obligate, intracellular parasites of higher plants, traditionally included with slime moulds as trophic organism is a small plasmodium; uninucleate spores release motile zoospores with 2 naked but heterodynamic flagella. 1 family.

Class Filosea

Naked and testate amoebae with thin hyaline pseudopodia which are sometimes branched and sometimes stiff, suggesting presence of microtubules; flagellated stages not known. Probably polyphyletic. 2 orders.

ORDER ACONCHULINIDA

Without external test, some with mucous sheath; some able to feed on algae by penetrating algal cell wall. 2 families.

ORDER GROMIIDA

Test present, often incorporating secreted siliceous elements; mostly from interstitial habitats. 6 families.

Subphylum Actinopoda

Amoebae with axopodia (pseudopodia with internal skeleton (axoneme) of cross-linked microtubules); typically with spherical symmetry, axopodia (arms) radiating from central body mass; mostly floating predators; sexual activity in some; flagellated cells produced in some; cytological organization quite variable. Many marine actinopods are very large and have inorganic skeletal elements; these have been referred to as the 'Radiolaria', an assemblage no longer regarded as a monophyletic taxon. The actinopoda are probably polyphyletic, with some groups having a closer affinity to some flagellates than to other amoebae. About 7000 species, almost one-half of which are fossils. 4 classes.

Class Acantharea

Usually with 10, 16, 20 or 32 simple or elaborated strontium sulphate spicules radiating from centre of body; all marine; outer vacuolated layer enclosed by sheath or envelope; towards cell centre, inner mass sometimes enclosed within fibrous capsule; often with large numbers of symbionts; outer layer varies in size in response to flotation requirements; cytological data sparse, nuclear division spindle located within nucleus; flagellated stages produced by some. 5 orders, 160 species.

ORDER HOLOCANTHIDA

Radial spicules fused in pairs at cell centre so 10 (rarely 16) spicules span diameter of cell; inner envelope lies far outside central mass of cytoplasm. 2 families.

ORDER SYMPHYACANTHIDA

With 20 spicules, typically fused into central mass of celestite (strontium sulphate); inner layer of envelope lies far outside central mass of cytoplasm. 3 families.

ORDER CHAUNACANTHIDA

With 20 spicules, rounded at internal ends and not firmly joined together; inner layer of envelope some distance outside central mass of cytoplasm. 3 families.

ORDER ARTHRACANTHIDA

With 20 radial spicules, tightly interconnected either by adpression of their tapered bases or by lateral expansions; inner layer of envelope closely adpressed to central mass of cytoplasm. 9 families.

ORDER ACTINELIIDA

With 16, 32 or many more radial spines. A poorly-known group with 3 families, including the only benthic acantharian.

Class Polycystinea

Marine, planktonic; with siliceous skeleton of solid elements in latticed shell with or without radial spicules; biflagellated 'isospores' produced by some; cytoplasm divided into inner and outer regions by central capsule perforated by many pores; microtubular supports of axopodia arise in central mass of cytoplasm; microtubules arranged in large polygonal arrays, or in branching lamellae derived from polygonal arrays; outer region of cytoplasm contains symbiotic algae (zooxanthellae). With the Phaeodarea (below), comprise the 'radiolarians'. 2 orders.

ORDER SPUMELLARIDA

Pores uniformly distributed in capsular membrane; skeleton absent or of spicules or of 1 or more concentric latticed shells. 12 families.

ORDER NASSELLARIDA

Capsular membrane with pores at one pole; skeleton formed from single element, often with conical shape. 9 families.

Class Phaeodarea

Skeleton, when present, contains organic matter in addition to silica; skeleton of hollow spicules or shells; inner mass of cytoplasm enclosed by capsule with cytostome at one pole and 2 openings at other side from which emanate pseudopodial axonemes; body usually contains a dark phaeodium, implicated in silica metabolism. Marine and planktonic; cytoplasm with many oil droplets and without algae. With the Polycystinea (above), comprise the 'radiolarians'. 6 orders.

ORDER PHAEOCYSTIDA

Skeleton absent or present only as spicules lying either free or arising from common junction point. 4 families.

ORDER PHAEOSPHAERIDA

Skeleton mainly in form of large latticed shell with polygonal meshes. 2 families.

ORDER PHAEOCALPIDA

Skeleton in form of small shell with numerous small pores and often with single larger opening; radial spines sometimes present. 4 families.

ORDER PHAEOGROMIDA

Skeleton in form of shell, sometimes reduced, sometimes incorporating diatom frustules; spines may be present. 2 families.

ORDER PHAEOCONCHIDA

Skeleton typically of 2 thick hemispherical valves pressed together. 1 family.

ORDER PHAEODENDRIDA

Skeleton formed from 2 valves, not closely adpressed, which may support long branching spicules. 1 family.

Class Heliozoea

Without elaborate inorganic skeletons although some may have scales or small external spicules; cytoplasm not, or not emphatically, divided into 2 zones; mostly freshwater. 4 orders.

ORDER DESMOTHORACIDA

Cell occupies perforated organic capsule attached to substratum by stalk; axopodia extend out of perforations; life-cycle includes flagellated and amoeboid stages; with kinetocyst-like extrusomes, microtubules of axonemes arranged in hexagons, and mitochondria with tubular cristae. Affinities uncertain. 1 family.

ORDER ACTINOPHRYIDA

Cells with 1 or many nuclei; microtubules of axonemes often arranged in double polygonal spiral; inner ends of microtubules often adpressed on a nucleus; with vesicular cristae in mitochondria, homogeneous osmiophilic extrusomes, diploid with autogamy in a cyst in most. One family characterized by presence of a flagellum. 2 families.

ORDER TAXOPODIDA

Microtubular axonemes insert on nucleus in 'ball and socket'-like articulations; contractile fibres associated with bases of arms cause axopodia to beat, rowing cell through water; siliceous spicules present, microtubules of axonemes in hexagonal arrays. 1 family.

ORDER CENTROHELIDA

Microtubules in axonemes disposed in hexagonal arrays, not terminating on nucleus but on central centroplast; organic or siliceous scales and/or spicules coat body of most; complex extrusomes; mitochondria with flattened cristae; some with cysts. Contains a diversity of forms, currently divided between 5 families, requiring further systematic attention.

Phylum Chrysophyta

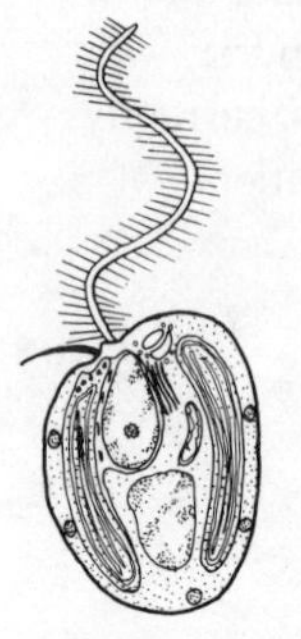

Flagellate, coccoid, amoeboid, filamentous or thalloid organisms, usually with a flagellate stage. The flagellates usually have two flagella, one (posterior) smooth and the other (anterior) bearing two rows of mastigonemes (thick hairs); flagella bases at right angles. The cell may be covered with large silica scales, with a delicate lorica, or may be naked. Two golden-brown chloroplasts are usually present, containing chlorophylls a, c_1 and c_2, β-carotene, fucoxanthin and other xanthophylls. Eyespot, when present, situated within chloroplast. The storage products are chrysolaminarin and fats. Some species phagotrophic. Mitochondria (with tubular cristae) and golgi present. Nuclear membrane usually breaks down, at least at spindle poles, in mitosis; spindle poles associated with rhizoplasts from flagellar bases. Probably haploid, sex involves fusion of flagellate isogametes. Produce siliceous cysts. Mainly found in freshwater. About 600 living and 200 fossil species in the single class, Chrysophyceae (some workers also include the Xanthophyceae, Bacillariophyceae, etc. in the Chrysophyta).

ORDER OCHROMONADALES

Cells with 2 unequal flagella inserted at anterior end; vegetative state, unicells or motile colonies. 6 families.

ORDER CHRYSAMOEBIDALES

Amoeboid phase dominant; in general, reproduction by flagellated stages. 5 families.

ORDER PHAEOTHAMNIALES

Form filamentous or parenchymatous thallus; reproduction by *Ochromonas*-like zoospores produced by any cell. 2 families.

ORDER CHRYSOSPHAERIALES

Main phase is coccoid cell, with reproduction by autospores or zoospores. 2 families.

ORDER PEDINELLALES

Unicells with only 1 emergent, ribbon-like flagellum; sometimes with stalk or peduncle and ring of 'tentacles' at anterior end. 1 family.

ORDER THALLOCHRYSIDALES

Form small prostrate parenchymatous base from which filaments may arise. 1 family.

ORDER HYDRURALES

Consist of much branched gelatinous 'skeleton' in which numerous cells embedded. 1 family.

ORDER DICTYOCHALES

Motile, naked unicells with numerous discoid chloroplasts and single anterior flagellum; silica skeleton of tubular elements formed internally; many known as fossils. 'Silicoflagellates.' 3 families.

Phylum Raphidophyta

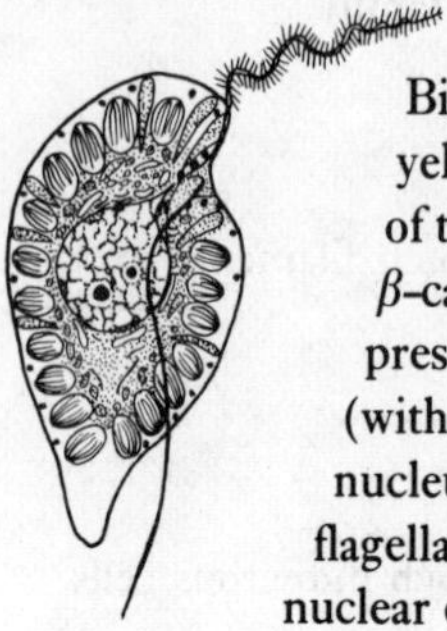

Biflagellated, naked unicells containing numerous yellow–green chloroplasts. The two flagella are like those of the Chrysophyta. Pigments include chlorophyll *a* and *c*, β-carotene and various xanthophylls. No eyespots are present. Oil is the main storage product. Mitochondria (with tubular cristae) and golgi are present. There is a large nucleus with characteristic rhizoplast connecting this to the flagellar bases. Trichocysts or mucocysts are present. The nuclear envelope has polar openings in mitosis, spindle microtubules originating from the flagellar bases. Sex has not been confirmed. Found in the sea and in freshwater. 'Chloromonads'. About 25 living species in a single order (the Vacuolariales), 1 family.

Phylum Bacillariophyta

The 'Diatoms' are single-celled organisms or colonies in which the cell wall is an elaborate silica box (frustule) of very varied shape. The cell wall is conserved, one-half to each daughter cell, following division. Each cell contains a single nucleus and usually several yellow–brown chloroplasts. Pigments present include chlorophylls a, c_1 and c_2, β-carotene and fucoxanthin. Food reserves are oil and chrysolaminarin. Mitochondria (with tubular cristae) and golgi present. Mitosis open without centrioles. Vegetative cells diploid, with meiosis in gamete formation. Resistant auxospores may be formed. Common in the sea and in freshwater as planktonic organisms, on mud surfaces and as epiphytes. Many known as fossils (Diatomite). About 10 000 living and probably 15 000 fossil species in a single class, the Bacillariophyceae.

ORDER BIDDULPHIALES

Basically radial organization and ornamentation; sexual reproduction involves spermatozoids with single anterior hairy flagellum and oogonia; mainly planktonic, occur singly or in chains. 11 families.

ORDER BACILLARIALES

Bilateral or asymmetric organization, often with pennate form; longitudinal grooves (or raphes) present which may be involved in gliding movement; sexual reproduction by a form of conjugation to give auxospores. 8 families.

Phylum Phaeophyta

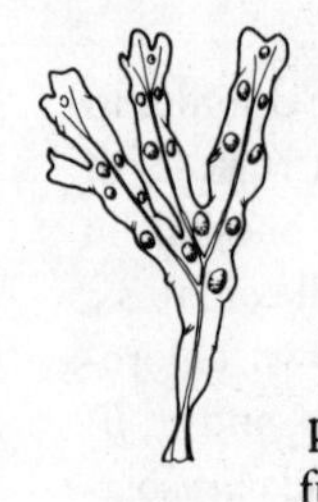

The 'Brown Algae' are multicellular, filamentous or thalloid marine forms, with a size range from less than 1 cm to 50 m, having a predominantly brown colour. Their cells are surrounded by a thick mucilaginous cell wall and contain yellow-brown chloroplasts. The main pigments are chlorophylls *a*, c_1 and c_2, β-carotene and a very large amount of fucoxanthin. The food reserve is laminarin. Mitochondria (with tubular cristae) and golgi are present. The nuclear membrane is open at the poles during mitosis, with centrioles at the spindle poles. Asexual reproduction is by biflagellate zoospores and sexual reproduction involves two biflagellate gametes or biflagellate spermatozoids and larger oogonia; the biflagellate cells have an anterior tinsel flagellum and a naked trailing flagellum. There is usually alternation of haploid gametophyte and diploid sporophyte generations, which may be isomorphic or heteromorphic. The main habitat is the intertidal belt, but the giant kelps are subtidal. They have commercial importance for extraction of alginates and as food. About 1500 species in a single class, the Phaeophyceae.

ORDER ECTOCARPALES

Relatively small, with filamentous or pseudoparenchymatous structure; reproduction isogamous or anisogamous; complex alternation of isomorphic generations, dependent on quality of light and temperature. 12 families.

ORDER CUTLERIALES

Trichothallic growth forms flattened blade in sporophyte phase and dissected thallus in gametophyte stage; sexual reproduction anisogamous. 1 family.

ORDER DESMARESTIALES

Trichothallic growth forms sporophyte thallus composed of large central cells and large numbers of small cortical cells; gametophyte stage, microscopic filament with oogamous reproduction. 2 families.

ORDER LAMINARIALES

Very large, parenchymatous; diploid sporophytic thallus alternates with microscopic haploid gametophytes with oogamous reproduction; sporophyte normally attached and (in giant kelps) may reach length of 50 m. 'Kelp', 'tangle', 'oar-weeds', etc. 4 families.

ORDER SPHACELARIALES

Branched, pseudoparenchymatous; growth governed by large apical cell; sexual reproduction isogamous; sporophyte and gametophyte generations isomorphic. 4 families.

ORDER DICTYOTALES

Flat, branched, thalloid alga with apical growth; sexual reproduction isogamous; sporophyte generation isomorphic with gametophyte. 1 family.

ORDER FUCALES

Growth of diploid parenchymatous gametophyte governed by recessed apical cell giving rise to flattened but branched thallus; sexual reproduction oogamous; no alternation of generations; dominate intertidal zone in many parts of world. 'Wracks', 'Shore-weeds', etc. 6 families.

Phylum Phycomycota

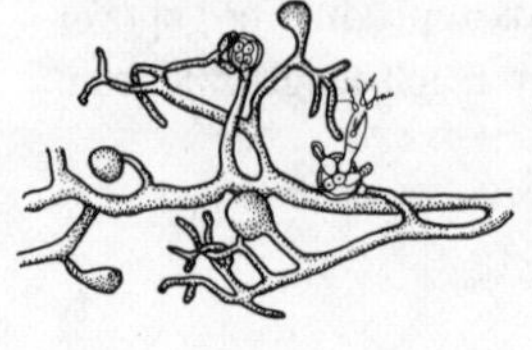

These protists were formerly classified as fungi (see p. 93), but give rise to asexual zoospores with an anteriorly-directed tinsel flagellum, and in one class also a posterior flagellum, the two basal bodies lying at 180°. The zoospores are formed in sporangia and germinate into simple structures that totally differentiate into a reproductive structure (holocarpic) or into a body of which part is reproductive while part remains somatic (eucarpic). The body usually extends into rhizoids or hyphae, with or without cross walls. Both body and hyphae are enclosed by a secreted wall of cellulose or chitin. They lack plastids but have both mitochondria (with tubular cristae) and golgi bodies. Centrioles are present outside the nucleus during mitosis, which may be closed or with polar fenestrae. Sexual reproduction is present in some forms; sexual forms are known to be diploid. They are parasitic or saprobic in freshwater or soil. About 800 species in 2 classes.

Class Oomycotea

Holocarpic or eucarpic; zoospores with anterior tinsel flagellum and posterior smooth flagellum; two types of zoospores may be produced in succession, a primary zoospore with both flagella originating at anterior end and a secondary zoospore with the two flagella originating laterally; zoospores germinate to produce a body with cellulose and glucan walls, with hyphae (where present) coarse and non-septate; nuclear division closed, with persistent nucleolus and intranuclear spindle with poles near pairs of centrioles oriented at 180° to each other; diploid with sexual reproduction by apposition of two hyphae cut off by septa at their tips, followed by migration of nuclei through channels from one (differentiated as an antheridium) to the other (differentiated as an oogonium), and formation of thick-walled (chitinous) oospores; oospores develop by release of zoospores; parasitic and free-living. 4 orders.

ORDER SAPROLEGNIIDA

Holocarpic or eucarpic with broad hyphae forming extensive mycelium; zoosporangia long and cylindrical; zoospores often dimorphic, sometimes only secondary zoospores present; oogonia usually contain several oospheres. 'Water moulds'. 5 families.

ORDER LEPTOMITIDA

Eucarpic with hyphae constricted at intervals with partial cellulin plugs; chitin sometimes present in walls; zoosporangia pyriform or short cylinders; 1 oosphere per oogonium; saprobic. 2 families.

ORDER LAGENIDIIDA

Small holocarpic body of 1 or few cells; zoosporangia may release a vesicle in which secondary zoospores develop; 1 oosphere per oogonium; mostly freshwater or marine parasites. 3 families.

ORDER PERONOSPORIDA

Eucarpic; zoosporangia usually globular, produced on special hyphae, detachable as wind-blown distributive phase; zoosporangia may function as conidia, germinating directly to produce germ tube, or may release secondary zoospores; 1 oosphere per oogonium; mostly terrestrial species in soil or on vascular plants; includes 'downy mildews'. 4 families.

Class Hyphochytridiomycotea

Holocarpic or simple eucarpic, with walls containing chitin and sometimes also cellulose; zoospores with only single anterior tinsel flagellum and barren basal body, formed within inoperculate sporangium or in plasmodial mass released from zoosporangium; nucleus at division perforated at poles and with centrioles in orthogonal pairs; gametes similar, flagellate; sexual or asexual resting spores that release zoospores on germination; parasitic or saprobic; freshwater and marine. 1 order, 3 families.

Phylum Labyrinthomorpha

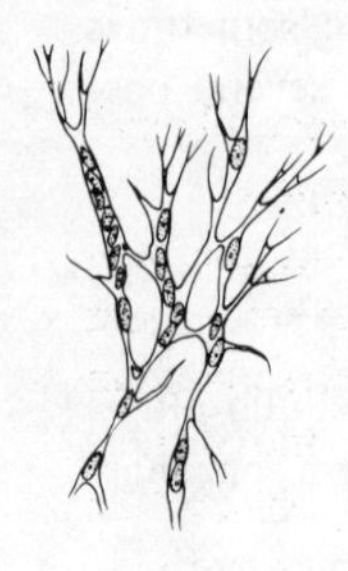

Cell produces hyaline cytoplasmic material in the form of a 'slime' or rhizoids, production of this material being associated with a sagenetosome, an organelle peculiar to this group; without chloroplasts; mainly marine, feeding on decaying or dead algae or as parasites of algae; zoospores produced with one tinsel flagellum and one naked flagellum; mitochondria with tubular cristae. 2 classes containing 25 species.

Class Labyrinthulea

Hyaline ectoplasm in form of slime channels within which spindle-shaped cells move; mainly from marine algae and grasses, sometimes as parasites; 1 freshwater species. 1 family.

Class Thraustochytridea

Parasitic; body mass on surface of decaying or living plant tissue, producing rhizoidal system which penetrates host tissue for nourishment. 1 family.

Phylum Xanthophyta

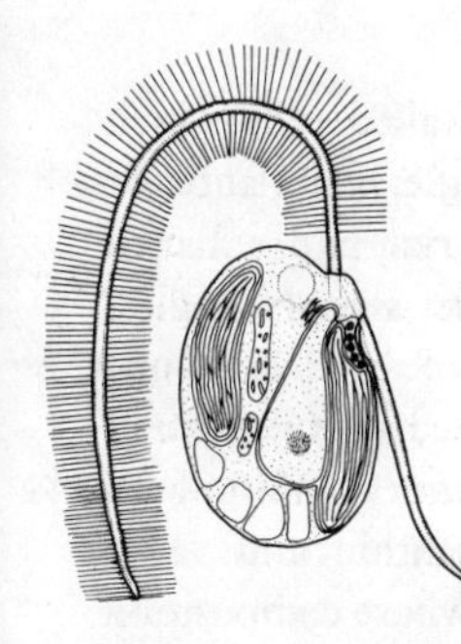

A group of filamentous, siphonous and coccoid algae which generally have biflagellate zoospores. The anterior flagellum bears mastigonemes and the posterior flagellum is smooth; their bases are set at right angles. The yellow-green chloroplasts contain chlorophylls *a*, c_1 and c_2, β-carotene and various xanthophylls. Eyespot situated within chloroplast; mitochondria (with tubular cristae) and golgi present. The food reserve material is oil or fat. Cell wall often composed of two halves. Haploid with biflagellate gametes; isogamous, anisogamous or oogamous. Nuclear division closed with intranuclear spindle and centrioles outside the nucleus, opposite the poles. Mainly freshwater organisms. About 650 species in a single class, the Xanthophyceae.

ORDER CHLAMYDOMYXALES

Amoeboid and palmelloid; flagellated reproductive stages. 11 families.

ORDER MISCHOCOCCALES

Coccoid; reproduction involves autospores and zoospores. 14 families.

ORDER TRIBONEMATALES

Filamentous, cellular organization; cell wall usually in 2 H-shaped pieces. 3 families.

ORDER VAUCHERIALES

Siphonaceous, filamentous, or bladder-like; multinucleate; reproduction by biflagellate or multiflagellate zoospores and sexual reproduction oogamous in *Vaucheria*. 1 family.

Phylum Eustigmatophyta

A small group consisting of coccoid, walled cells which reproduce by zoospores which have a single, hairy, anterior flagellum or, occasionally, a hairy anterior and a naked posterior flagellum, with the two bases set at right angles. The composition of the cell wall, which appears to be in one piece, is unknown. A large eyespot is situated in cytoplasm at the anterior end of the cell. The yellow-green chloroplasts contain chlorophyll *a*, β-carotene, violaxanthin and vaucheriaxanthin. The food reserve forms lamellate bodies whose composition has not been determined. Mitochondria (with tubular cristae) and golgi bodies are present. Nuclear division has not been described and no sexual reproduction is known. Mainly freshwater organisms but also found in soil and marine plankton. About 10 species in a single order (the Eustigmatales), 4 families.

Phylum Haptophyta

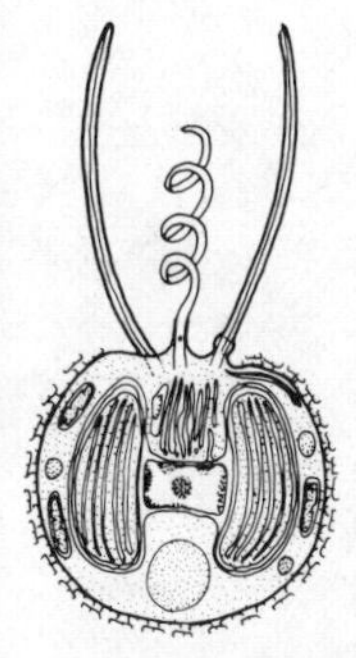

Small, single-celled organisms which, when in the motile state, have two non-hairy flagella with bases set at right angles and a third filiform appendage, the haptonema, which is often coiled and has a characteristic structure including a core of six or seven single microtubules. In some cases the haptonema is reduced to a small stub. A few species have a chloroplast-associated eyespot. In general, the cells contain one nucleus and two yellow chloroplasts. Pigments present include chlorophylls *a*, c_1 and c_2, β-carotene, fucoxanthin and various other xanthophylls. Food reserve product is chrysolaminarin. Phagotrophy reported in some species. Mitochondria (with tubular cristae) and golgi present. The nuclear membrane breaks down during mitosis and the spindle poles are near flagellar basal bodies. Haploid and diploid phases alternate in at least some species, the haploid form producing flagellate isogametes. Many members of this group bear organic or calcareous scales (coccoliths) and are known as 'coccolithophores'. Mainly marine organisms. About 450 living and 1100 fossil species in a single class, the Prymnesiophyceae.

ORDER ISOCHRYSIDALES

Haptonema rudimentary or very small; cell naked, lacks scales. 4 families.

ORDER COCCOSPHAERALES

Haptonema much reduced; cells normally covered with scales, some delicate and organic, others calcareous (coccoliths); some with complex life history involving benthic phases. 9 families.

ORDER PRYMNESIALES

Haptonema distinct, may be very long; cells covered with elaborate organic scales. 2 families.

ORDER PAVLOVALES

Short haptonema present; flagella distinctly unequal in length, covered with small club-shaped scales. 1 family.

Phylum Cryptophyta

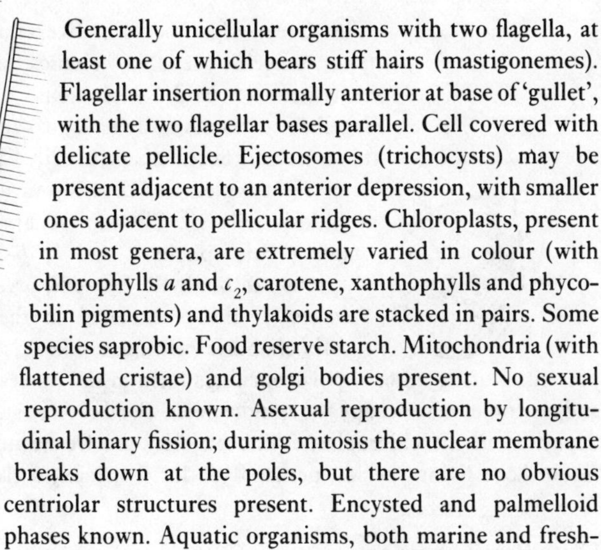

Generally unicellular organisms with two flagella, at least one of which bears stiff hairs (mastigonemes). Flagellar insertion normally anterior at base of 'gullet', with the two flagellar bases parallel. Cell covered with delicate pellicle. Ejectosomes (trichocysts) may be present adjacent to an anterior depression, with smaller ones adjacent to pellicular ridges. Chloroplasts, present in most genera, are extremely varied in colour (with chlorophylls *a* and c_2, carotene, xanthophylls and phycobilin pigments) and thylakoids are stacked in pairs. Some species saprobic. Food reserve starch. Mitochondria (with flattened cristae) and golgi bodies present. No sexual reproduction known. Asexual reproduction by longitudinal binary fission; during mitosis the nuclear membrane breaks down at the poles, but there are no obvious centriolar structures present. Encysted and palmelloid phases known. Aquatic organisms, both marine and freshwater. About 200 species in a single class, the Cryptophyceae.

ORDER CRYPTOMONADALES

Vegetative cells flagellated; motile cells covered with delicate periplast; some have chloroplasts, others with saprobic nutrition. 6 families.

ORDER TETRAGONIDALES

Tetrahedral, coccoidal, walled cells in vegetative phase reproducing by cryptomonad-like zoospores. 1 family.

Phylum Chlorophyta

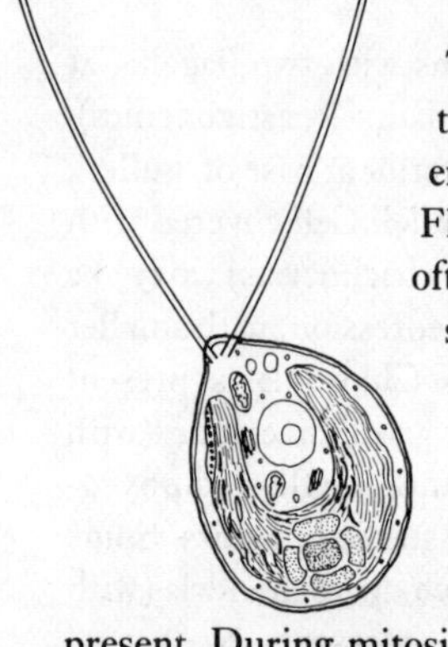

The 'Green Algae' occur in diverse morphological types: unicells, coenobia, colonies, filaments, parenchymatous and siphonous organization. Flagellate stages normally have two anterior flagella often with hairs or scales, and usually with their bases set at right angles. Cell covering varied but often a cellulosic cell wall. Chloroplasts grass-green containing chlorophylls *a* and *b*, *β*-carotene and various xanthophylls. Eyespot, present in motile stages, is always enclosed in the chloroplast. Mitochondria (with plate-like cristae) and golgi present. During mitosis, polar openings form in the nuclear membrane, and centrioles may occupy the spindle poles. Food reserve starch, which is deposited in the chloroplasts. The majority are freshwater organisms. About 8000 living species, perhaps 500 fossil species, 6 classes.

Class Prasinophyceae

Generally only occur as flagellated unicells; variable number of flagella, generally covered with scales; cell bodies scale-covered in some; flagellar roots broad and contact the nucleus; sexual reproduction not confirmed.

ORDER PEDINOMONADALES

With only 1 flagellum and 1 chloroplast; very small cells. 1 family.

ORDER PTEROSPERMATALES

Free-floating, complex 'cysts' with thick wall extended into a 'wing' and containing numerous chloroplasts; reproduce by autospore formation. 2 families.

ORDER HALOSPHAERALES

Spherical, multinucleate, planktonic cells which produce large numbers of zoospores in reproduction. 1 family.

ORDER PYRAMIMONADALES

Vegetative stage in form of flagellated unicells with scale-covered body as well as scaly flagella. 1 family.

ORDER PRASINOCLADIALES

Main form is a small multicellular benthic alga; reproduction by quadriflagellate zoospores. 1 family.

Class Charophyceae

Macroscopic; composed of extremely large cells forming rhizoids and stems with whorls of branches at nodes; growth takes place from an apical cell; cell walls may be heavily calcified; sexual reproduction by flagellated antherozoids and oogonia produced in complex oogonium; vegetative stage haploid with meiosis in division of zygote; mainly freshwater. Many fossil forms. A single order (Charales), 1 family.

Class Chlorophyceae

Flagella non-scaly; flagellar roots run in periphery of cell; usually haploid, but alternation of haploid and diploid generations in some; morphology various.

ORDER VOLVOCALES

Vegetative stages flagellates or coenobia; zoospores and gametes normally biflagellate. 2 families.

ORDER TETRASPORALES

Colonial; no motile stages. 1 family.

ORDER PRASIOLALES

Vegetative stage small, multicellular thallus; reproduction by flagellated zoospores and gametes. 1 family.

ORDER CHLOROCOCCALES

Vegetative stage non-motile unicells or complex colonies; asexual reproduction by zoospores or autospores; sexual reproduction by motile gametes. 7 families.

ORDER ULOTRICHALES

Uniseriate filaments of uninucleate cells; asexual and sexual reproduction by flagellated cells. 3 families.

ORDER ULVALES

Parenchymatous seaweeds reproducing by flagellated zoospores and gametes. 2 families.

ORDER CHAETOPHORALES

Branched filaments with flagellated reproductive stages; vegetative stage often trichothallic. 2 families.

Class Oedogoniophyceae

Filamentous, with uninucleate cells and complex method of cell division ('cap cells'); haploid, with meiosis in division of zygote; flagellated zoospores and male gametes bear ring of numerous flagella. A single order (Oedogoniales), 1 family.

Class Bryopsidophyceae

Multinucleate or siphonous; morphology varied.

ORDER CLADOPHORALES

Branched filaments of multinucleate cells; reproduction by flagellated zoospores and gametes; alternation of haploid gametophyte and diploid sporophyte generations; cell wall thick and rough; marine and freshwater. 1 family.

ORDER BRYOPSIDALES

Branched siphonous (non-cellular) organization; diploid, with meiosis in gamete formation; reproduction by flagellated cells; marine. 2 families.

ORDER CODIALES

Multiaxial siphonous construction forming macroscopic seaweeds; reproduction mainly by flagellated stages but female gamete sessile; probably all diploid. 6 families.

ORDER DASYCLADALES

Thallus complex with radial symmetry, sometimes of bladder-like or umbrella-like form; 1 nucleus or many; cell wall of cellulose or may become calcified; diploid. 4 families.

Class Conjugatophyceae

Filamentous or unicellular; reproduction involving conjugation between 2 cells; no flagellated stages; haploid, with meiosis in division of zygote; freshwater.

ORDER DESMIDIALES

Variously shaped unicells; usually 2 chloroplasts with nucleus sited between them, often in narrow isthmus. 'Desmids'. 2 families.

ORDER ZYGNEMATALES

Unbranched, filamentous; varied forms of chloroplasts; cell wall usually slimy; conjugation results in resistant zygospore. 1 family.

Phylum Chytridiomycota

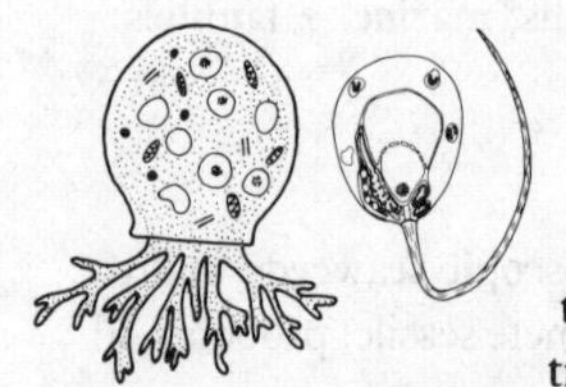

These microscopic protists may be formed entirely of a reproductive body (holocarpic) or have branched, tapering rhizoids extending from the reproductive body (eucarpic). They lack plastids but possess mitochondria (with flattened cristae) and golgi bodies. The mitotic spindle is internal with paired polar centrioles arranged orthogonally outside a closed nucleus (Blastocladiidea) or one with polar fenestrae (Chytridiidea). Asexual zoospores, which are released from sporangia and propelled by a single posterior flagellum, usually have a concentration of ribosomes adjacent to the nucleus, prominent oil droplet(s) and a rumposome body (a membranous body near the cell surface and associated with microbodies, microtubules, one or more lipid droplets and often one or more mitochondria). Sexual reproduction may be isogamous or anisogamous, at least one of the gametes usually being motile and like a zoospore, producing a zygote which may germinate directly or form a resting spore. The secreted body wall is of chitin and glucan. They are parasitic or saprobic. About 900 species in 2 classes.

Class Chytridiidea

Holocarpic or eucarpic; zoospore with a rumposome, a barren basal body parallel to and attached to side of active basal body, with microtubules extending from one side of active basal body to rumposome and/or to other parts of cytoplasm but not enclosing the nucleus, and with ribosomes aggregated near nucleus and associated with membranous bodies but not forming a distinct nuclear cap; main part of life-cycle haploid. 2 orders.

ORDER CHYTRIDIIDA

Holocarpic or eucarpic with rhizoidal system; zoospore with anterior nucleus, single large lipid droplet, single mitochondrion around flagellar base, and microtubules in bundle connecting active basal body to rumposome; zoosporangium spherical or pyriform; germination monopolar; gametes variable. 11 families.

ORDER MONOBLEPHARIDIIDA

Eucarpic with filamentous thallus; zoospore with central nucleus, many small lipid bodies, many mitochondria, and microtubules radiating from active basal body to cytoplasm but not to rumposome; zoosporangia cylindrical or pyriform; germination bipolar; sexual reproduction oogamous with non-flagellate female gamete; resting oospore. 2 families.

Class Blastocladiidea

Holocarpic or eucarpic; zoospore without rumposome (though a side-body, present in some, may be a reduced analogue), but with barren basal body not parallel to or attached to side of active basal body; ribosomes either dispersed or aggregated into prominent nuclear cap; microtubules or striated root arising by or at proximal end of active basal body extend to and often around nucleus, which is close to basal bodies; alternation of haploid and diploid phases probably usual; male and female gametes flagellate, but may be of different size. 2 orders.

ORDER BLASTOCLADIIDA

Eucarpic, with basal cell anchored by branched tapering rhizoids; produce thin-walled and thick-walled pigmented zoosporangia; zoospores with prominent nuclear cap enclosed by double envelope connected to nuclear membrane and with microtubules that extend from proximal end of active basal body to surround nucleus; germination bipolar. 3 families.

ORDER SPIZELLOMYCETIDA

Holocarpic; zoospore with ribosomes dispersed and basal bodies usually connected to nucleus by striated rhizoplast; zygote forms resting sporangium; germination monopolar. 2 families.

Phylum Choanoflagellata

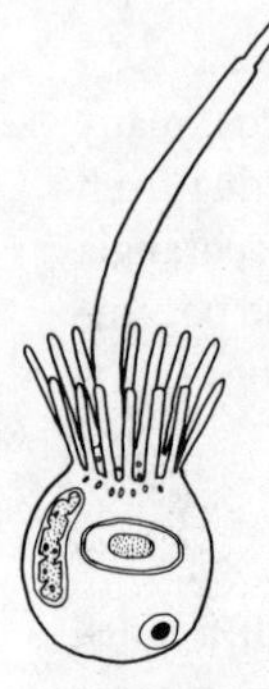

These are free-living, solitary or colonial small unicells thought to have affinities with sponges. They bear a single apical flagellum, the basal part of which is surrounded by a ring (collar) of tentacular microvilli used in filter feeding. Microtubular flagellar rootlets radiate from all around the active basal body and terminate beneath the cell surface membrane, without making any connection with the nucleus or with the barren orthogonal basal body. There are no striated filamentous roots. The phagotrophic cells lack plastids, but mitochondria (with flat cristae) and golgi bodies are present. They usually secrete a membranous sheath or basket-like lorica of siliceous bars. Stalked and free-swimming. Details of nuclear division not available. About 140 species in a single order containing a single family.

Phylum Rhodophyta

The 'Red Algae' have unicellular, filamentous, pseudoparenchymatous or parenchymatous construction. No flagella are found in this group. Cell covering is a cell wall consisting of microfibrils (xylan, mannan) and much gelatinous material including agar; some are calcified. Most contain red chloroplasts with a distinctive ultrastructure in which the thylakoids are separate, with chlorophyll *a* (and sometimes *d*), β-carotene, other xanthophylls and red and blue phycobilin pigments. Food reserve is floridean starch which forms in the cytoplasm. Mitochondria (generally with plate-like cristae) and golgi are present. Nuclear membrane with polar openings in mitosis, no centrioles. Most are marine. The group has a long fossil history. About 5000 species (of which 750 are fossils) in 2 classes.

Class Bangiophyceae

Relatively simple, consisting of unicells, filaments and parenchymatous forms; spores formed by division of vegetative cell into 4.

ORDER PORPHYRIDIALES

Mostly unicells, with no known method of sexual reproduction; multicellular forms haploid, forming spermatia and simple carpogonia; zygote undergoes meiosis during carpospore formation; some can move by gliding. 2 families.

ORDER BANGIALES

Alternation of generations between thalloid stage and a filamentous stage (conchocelis) inhabiting old mollusc shells; many types of spores formed; sexual reproduction probably occurs. 3 families.

Class Florideophyceae

Filamentous or pseudoparenchymatous; sexual reproduction by spermatia and very complex carpogonium leading to formation of carpospores; in most, an alternate sporophyte generation produces tetraspores; gametophyte stage haploid, carposporophyte stage either haploid or diploid; tetrasporophytes diploid with meiosis in tetraspore formation; distinct pit connections occur between cells. Taxonomy generally based on details of the complex sexual reproductive apparatus.

ORDER NEMALIONALES

Multiaxial filamentous construction forming rather gelatinous algae. 12 families.

ORDER GIGARTINALES

Small, with multiaxial pseudoparenchymatous construction and rather cartilaginous texture. 23 families.

ORDER CRYPTONEMIALES

Multiaxial thalli and coralline algae. 15 families.

ORDER PALMARIALES

Single genus, with edible, flattened, pseudoparenchymatous thallus ('dulse').

ORDER RHODYMENIALES

Thallus, uniaxial, pseudoparenchymatous; small seaweeds. 2 families.

ORDER CERAMIALES

Mostly branched, filamentous; some form delicate thalli by pseudoparenchymatous aggregation and by down-growth of corticating rhizoidal filaments. 4 families.

Phylum Microspora

These parasites produce unicellular spores enclosed by an imperforate wall within which one uninucleate or dinucleate sporoplasm is associated with an extrusion apparatus composed of a polar cap and an eversible polar tube; the amoeboid sporoplasm emerges from the everted polar tube when the spore hatches. This often develops into a syncytial plasmodium. They possess no plastids, mitochondria or flagella, but golgi bodies are present. Mitotic divisions are closed with an intranuclear spindle whose poles centre on plaques (and sometimes centrioles ?). Sexual stages suspected but not proven. Spores may be formed in large numbers in large parasitophorous vacuoles enclosed by a single membrane, or in smaller numbers within a thick sporocyst wall inside a vacuole. They are found as intracellular parasites in animals of almost all major groups and in protists. About 800 species in 2 classes.

Class Rudimicrosporea

Hyperparasites of gregarine apicomplexans in annelids; spores, spherical or subspherical and simple, with polar cap and short thick polar tube but no polaroplast or posterior vacuole; sporocysts and parasitophorous vacuoles produced by same parasite within same host. A single order (Metchnikovellida), 1 family.

Class Microsporea

Spores with complex extrusion apparatus, usually with polaroplast around attachment of polar tube to polar cap, as well as posterior vacuole; polar tube typically long, coiled around nucleus within spore; spore wall of 3 layers, with proteinaceous exospore, chitinous endospore and inner membranous layer; sporocysts present or absent.

ORDER MINISPORIDA

Spores relatively simple, both polaroplast and endospore little developed and polar filament short; sporocyst usually well developed; spores usually spherical or ovoid. 3 families.

ORDER MICROSPORIDA

Spores complex, with well developed extrusion apparatus and spore wall; sporocysts reduced, sometimes absent; spore shape variable. 13 families.

Phylum Ascetospora

These parasites of invertebrates have complex unicellular or multicellular spores which differentiate within a plasmodial mass. The spores lack polar capsules or polar filaments and contain one or more sporoplasms. One or more of the spore cells contains haplosporosomes, dense bodies (about 0.2 μm × 0.3 μm) enclosed in one or two membranes. The cells lack plastids and flagella, but possess mitochondria (with flattened cristae) and golgi bodies. Mitotic divisions are closed with an intranuclear spindle and pairs of orthogonally arranged centrioles outside the nuclear envelope at either pole of the spindle. About 30 species in 2 classes.

Class Stellatosporea

Spores unicellular, with apical opening covered by cap or underlain by diaphragm. A single order (Balanosporida), 1 family.

Class Paramyxea

Spores multicellular, produced by series of concentric endogenous divisions, producing cells within cells; spores without opening. A single order (Paramyxida), 2 families.

Phylum Myxozoa

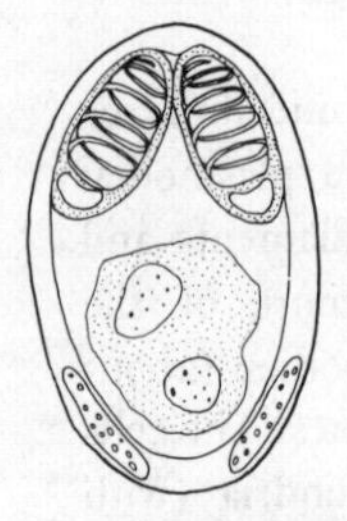

Parasites which grow as plasmodia within animals. Portions of the plasmodium become separated by membranes and differentiate to produce spores. Spores are of multicellular origin with one or more polar capsules and sporoplasms, and with a wall formed of one to three (or rarely more) valves. Each polar capsule contains a coiled filament, capable of eversion, probably for anchorage, prior to dehiscence of the spore. They contain mitochondria with plate-like cristae, but neither plastids nor flagella are present. Nuclear division is closed with an internal spindle whose poles are centred on plaques at the nuclear membrane. They are believed to be diploid. About 900 species in 2 classes.

Class Myxosporea

Spores with 1–2 sporoplasms and usually 2 (occasionally up to 6) polar capsules; usually 2 valves in spore wall (occasionally up to 6); trophozoite stage well developed and main increase in numbers occurs in this stage; in body spaces and tissues of ectothermic vertebrates.

ORDER BIVALVULIDA

Spore walls with 2 valves; usually 2 (occasionally 4) polar capsules at same or opposite poles of spore. 7 families.

ORDER MULTIVALVULIDA

Spore walls with 3 or more valves. 3 families.

Class Actinosporea

Spores with 3 polar capsules and spore wall of 3 valves; several or many sporoplasms; main increase in numbers occurs in sporogenesis and trophozoite stage reduced; in invertebrates, principally annelids. A single order (Actinomyxida), 4 families.

Protists of Uncertain Affinity

Protista whose affinities remain to be resolved include the following 3 orders.

ORDER BICOSOECIDA

Small flagellates with 2 flagella originating from anterior depression; longer anterior flagellum with stiff hairs; shorter smooth flagellum runs in groove across cell surface and turns back to anchor cell within base of secreted cup-shaped lorica; without plastids; with golgi body and mitochondria (tubular cristae); phagotrophic; among flagellar roots is broad band of microtubules running around lip of cytostome, but no rhizoplast; freshwater and marine. 30–40 species in a single family. Sometimes classed with chrysophytes.

ORDER EBRIIDA

Unicellular biflagellates which may be related to dinoflagellates; without plastids, but with fat droplets and complex internal silica skeleton; phagotrophic members of marine plankton. 2 living genera in separate families and many fossil representatives. Ultrastructural study required.

ORDER PROTEROMONADIDA

Parasitic flagellates with 1–2 pairs of heterodynamic flagella; basal bodies of each pair set orthogonally; flagella arise near anterior end of body, without a flagellar pocket, without paraxial rods (with slender paraxial strand in 1 genus); microtubular flagellar roots either join pellicular microtubules (1 genus) or are joined by dense and striated fibres to form a rhizoplast associated with both nucleus and large postnuclear mitochondrion (with tubular cristae but no kinetoplast); ridges of body surface contain microtubules, pinocytotic digestive vacuoles form between ridges; without plastids, but with single large anterior golgi body and glycogen granules. 2 families and perhaps 6 species; sometimes classed as kinetoplastids, but more likely related to opalinids.

There are a number of other flagellates, all lacking chloroplasts, whose distinctive structures have been studied in sufficient detail to show that they cannot be assigned satisfactorily to any existing order of Protista as defined here. They may come to be regarded as small independent groups of ordinal or even higher rank. These include *Apusomonas*, *Cercobodo* (= *Cercomonas*), *Colponema*, *Cyanophora* (with cyanobacterial symbiont), *Cyathobodo*, *Rhipidodendron* (and *Spongomonas*).

KINGDOM FUNGI

H. J. HUDSON

Any definition of a fungus must be very imprecise simply because of the immense variety of eukaryote organisms usually included as fungi. They occur free-living or as parasitic or mutualistic symbionts. All are heterotrophic for carbon compounds, as unlike plants and many protists they lack chlorophyll and they, the free-living ones at least, secrete extracellular lytic enzymes and absorb soluble nutrients rather than ingest solids or liquids as animals do. They range in size from microscopic to massive. They may consist of a single ellipsoidal cell as in many true yeasts and the yeast-like phase of other fungi, but the basic structural unit in by far the majority is the hypha. This is a walled, tubular, much-branched filament, which may or may not possess cross walls or septa. The septa may be of a variety of types: complete; with a central simple pore; or with a flanged and capped central pore (dolipore septum). Hyphae extend by apical growth, only the very tip being capable of extension. Branches arise profusely behind the tip to form the vegetative phase or mycelium.

Gross hyphal morphology is of little diagnostic value but the presence or absence of septa, their type and the nature of their cell walls are important at the phylum or subdivision level. Hyphae differ in both wall composition and structure. The walls of most hyphae, certainly all septate ones, contain chitin, a polymer of β1.4-linked 2,acetamido-2-deoxy-D-glucopyranose. It is usually associated with non-cellulosic β1.6-linked glucans. True cellulose, characteristic of plants, is usually lacking.

Hyphae vary in diameter but the majority fall within the range 1–15 μm. Each compartment of a septate hypha may contain from one to many nuclei. The nuclear phases differ amongst fungi and from other organisms. In all Zygomycotina and Ascomycotina, the vegetative phase is haploid: it is a haplophase. At sexual reproduction, two haploid nuclei fuse to produce the diplophase which in many, such as in the Zygomycotina, is restricted to a single cell, the zygospore. In the Ascomycotina, the essential part of the sexual process is initiated by the fusion of two cells (plasmogamy) containing one or more nuclei. Nuclear fusion (karyogamy) does not occur immediately but the two nuclei form a pair—a dikaryon. This pair of closely associated, compatible nuclei then divide together to produce sister nuclei which are separated into two daughter cells. Such repeated divisions establish a dikaryophase. In the Ascomycotina, this dikaryophase is purely reproductive. It is restricted to the hyphae which produce asci (the ascogenous hyphae). However, in the Basidiomycotina, it is the haplophase which is restricted, and the vegetative phase is the dikaryophase. Nuclear fusion and meiosis occur in the basidium initial.

Such a dikaryophase is unique to these fungi and is functionally diploid, but for the Basidiomycotina, at least, it would appear to have a selective advantage over the true diploid.

Fungi reproduce both asexually and sexually by spores. These may be borne free on the vegetative mycelium or enclosed within special reproductive structures, such as ascocarps or basidiocarps in the Ascomycotina and Basidiomycotina respectively. Here their life styles are unique. Many may show two separate but contemporaneous phenotypes. The holomorph or whole organism in the fungi often consists of: (a) a teleomorphic state reproducing sexually by producing perfect spores, such as ascospores or basidiospores, as a result of a nuclear fusion followed by a meiosis; and (b) an anamorphic state reproducing asexually by producing imperfect spores, such as conidia, as a result of mitotic divisions. Each state thus produces a different type of spore. In the Deuteromycotina or 'Fungi Imperfecti', the anamorphic state appears, at least, to exist independently.

Traditionally, two Divisions have been recognized in the fungi—the Myxomycota and the Eumycota. The former are wall-less and quite unusual organisms, only included in the fungi as they have been mostly studied by mycologists. They possess either a plasmodium, a mass of naked, multinucleate protoplasm, which feeds by ingesting particulate matter and moves by amoeboid movement, or a pseudoplasmodium, an aggregation of separate amoeboid cells. Both of these forms are of a slimy consistency, hence their common name 'slime moulds'. Three major groups have been recognized: the Acrasiomycetes, the cellular slime moulds; the Myxomycetes, the true slime moulds; and the Plasmodiophoromycetes, the endoparasitic slime moulds. They are considered here as part of the Protista (pp. 52–53 and p. 54).

In the Eumycota or true fungi, it has been customary to recognize five subdivisions, equivalent to phyla, based primarily on the nature of the spores formed, especially the sexually-produced ones. These are the Mastigomycotina (= Phycomycotina), Zygomycotina, Ascomycotina, Deuteromycotina and Basidiomycotina. The Mastigomycotina, unlike other true fungi, all produce motile flagellate stages in their life-cycle, be these zoospores or motile gametes. Two types of flagella occur: the smooth whiplash and the tinsel with lateral hairs, flimmers or mastigonemes borne singly down both sides of the main axis. Three distinct types of zoospores are produced: those with a single posterior whiplash flagellum; those with a single anterior tinsel flagellum; and those with two subapically or laterally attached flagella, an anteriorly directed tinsel and a posteriorly directed whiplash. This suggests that the group is polyphyletic being merely an assemblage of organisms which only resemble each other to the extent that they reproduce asexually by zoospores. Three classes have been erected in the Mastigomycotina: the Chytridiomycetes, the Hyphochytridiomycetes and the Oomycetes. Members of the former class all possess only one common feature—a zoospore with a single posterior flagellum. All the Hyphochytridiomycetes possess zoospores with a single anterior tinsel flagellum and the Oomycetes possess biflagellate zoospores. The Oomycetes are so-called because they reproduce oogamously like some algae. They differ so markedly from all other fungi, except in their modes of nutrition, that most authors now accept that they are heterotrophic, coenocytic oogamous algae. They are here classified as such and treated in the section on the Kingdom Protista (p. 66). The Hyphochytridiomycetes are a small and difficult group, but many authors now consider them to be more closely related to the algae than to the fungi. They are also treated here as Protista (p. 67). The Chytridiomycetes are more problematical. They are certainly like fungi in a number of respects; for example, in their mode of nutrition and the fact that they have chitin in their walls and no cellulose. Members of the other two classes are like plants in having some cellulose in their walls. As no members of the other four subdivisions of true fungi possess zoospores at any stage of their life-cycles, it would seem most appropriate that the chytrids should also be included in the Protista with the other flagellate forms (p. 78).

It is becoming abundantly clear that the Deuteromycotina are also a most artificial assemblage, consisting of forms related to the Ascomycotina in the main, although some are related to the Basidiomycotina. They are fungi which have lost the ability to reproduce sexually or are forms whose sexual state has not yet been discovered or connected to the asexual conidial state. At present it would seem advisable to retain the subdivision, if only for convenience.

Phylum (Subdivision) Zygomycotina

All those fungi which reproduce asexually by non-motile sporangiospores (aplanospores) or by modified sporangia functioning as conidia, and sexually by the complete fusion of two, usually morphologically similar, gametangia resulting in the formation of warty, thick-walled, resting zygospores. Vegetative hyphae are haplophase and aseptate. Septa only delimit reproductive structures and dead or injured parts. Cell walls are of chitin and chitosan. The 610 species are apportioned between 2 classes and 7 orders.

Class Zygomycetes

One of the more natural groups of fungi because of the production of asexual aplanospores, fusion of gametangia to produce zygospores and walls of chitin and chitosan; asexual evolution appears to be from many-spored sporangia, through sporangia with a much-reduced number of spores, to one-spored sporangiola which function as conidia.

ORDER MUCORALES

Aplanospores produced in globose, multinucleate sporangia, narrow, cylindrical, sac-like merosporangia, in few-spored sporangiola or singly as conidia; zygospores often thick-walled, black and warty resting spores; large terminal chlamydospores common in mycorrhizal forms. Saprotrophic 'pin moulds' but also including a number of biotrophic parasites of other Mucorales and the fungi of vesicular–arbuscular mycorrhiza. 14 families.

ORDER ENTOMOPHTHORALES

Some saprotrophic, but mostly insect parasites; vegetative phase tending to break up into segments (hyphal bodies); asexual reproduction by forcibly discharged uni- or multinucleate conidia; zygospores smooth or ornamented. 1 family.

ORDER ZOOPAGALES

Majority predacious on rhizopod protists, or on free-living nematodes, or endoparasitic in them; vegetative phase of branched, aseptate hyphae or coiled, branched or unbranched filament; asexual reproduction by conidia; zygospores covered with hemispherical warts. 2 families.

Class Trichomycetes

A group of uncertain affinities. Mostly obligate symbionts of arthropods, attached to chitinous gut linings, especially of hindgut, or, externally, to exoskeleton. Vegetative phase much reduced, unbranched and coenocytic, or branched, septate and attached by basal cell; asexual reproduction by exogenous spores with appendages (trichospores), arthrospores, sporangiospores or even amoeboid cells; resting spores (zygospores) produced by some, by fusion of protoplasts.

ORDER HARPELLALES

Branched or unbranched vegetative phase attached to lining of hindgut or to peritrophic membrane of immature aquatic insects; trichospores with 1 to many basally-attached, fine appendages; some produce biconical zygospores. 2 families.

ORDER ASELLARIALES

Branched vegetative phase attached to lining of hindgut of insects or isopods; arthrospores produced by hyphal fragmentation; sexual reproduction unknown. 1 family.

ORDER ECCRINALES

Vegetative phase unbranched or branched only at base, attached to chitinous lining of hindgut of wide variety of arthropods; sporangiospores produced singly in series of terminal sporangia releasing spores in sequence; sexual reproduction unknown. 3 families.

ORDER AMOEBIALES

Unbranched vegetative phase in hindgut or on exoskeleton of aquatic crustaceans and insects; entire phase becomes converted into sporangium releasing amoeboid motile cells or walled spores; sexual reproduction unknown. (Relationship to other Trichomycetes has been questioned). 1 family.

Phylum (Subdivision) Ascomycotina

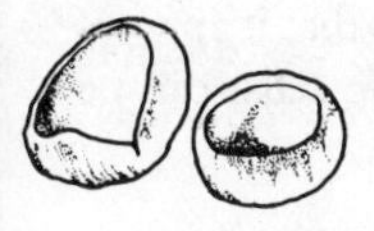

Vegetative haplophase. Hyphae septate, with simple pore. Ascospores produced within an ascus and often enclosed in an ascocarp. Nuclear fusion in ascus followed by meiosis and usually a mitosis and the formation of 8 ascospores. Asexual conidia often present. Wall of chitin and glucan.

The 15 000 species are apportioned between 6 classes and 20 orders, and also included is the Lecanorales (lichens) which contains a further 18 000 species.

Class Hemiascomycetes

Vegetative phase unicellular yeast-like or filamentous; asci one-walled, naked, not borne on ascogenous hyphae, produced singly, following karyogamy; no ascocarps.

ORDER ENDOMYCETALES

Unicellular or filamentous; asexual reproduction by budding or fission; sexual reproduction by fusion of 2 cells, the product directly, or after division, forming the ascus. True 'yeasts'. 4 families.

ORDER PROTOMYCETALES

Biotrophic parasites of seed plants causing lesions and galls often with marked changes in colour. Intercellular mycelium diploid, developing into thick-walled resting spores; resting spores germinate by rupture of outer wall and protrusion of cell membrane, or by papilla-like extension of outer wall to form enlarged multinucleate spore sac (synascus), regarded as equivalent of several asci; nuclei undergo meiosis and eventually form mass of endospores, forcibly discharged *en masse*; endospores bud and eventually fuse to form diploid mycelium prior to infection. 1 family.

ORDER TAPHRINALES

Biotrophic parasites of seed plants and ferns causing galls, leaf curls, deformed fruits and witches brooms. Intercellular or subcuticular dikaryophase mycelium in parasitic phase with terminal chlamydospores or ascogenous cells, each forming single ascus in hymenium-like layer; after nuclear fusion and mitosis, ascogenous cells often divide to give basal stalk cell with ascus at apex; ascospores may bud within ascus so that it appears multispored; saprotrophic phase of budding monokaryotic cells; can be cultured in this yeast state. 1 family.

Class Plectomycetes

Asci one-walled, evanescent, more or less globose, borne at all levels within ascocarp and arising on ascogenous hyphae, within closed, globose ascocarp (cleistothecium), not accompanied by packing paraphyses.

ORDER EUROTIALES

Mostly saprotrophic in soil and on variety of plant and animal debris, e.g., dung, feathers, horn and hooves. Ascocarps variable in form but usually globose; asci globose or subglobose, without stalk, usually 8-spored, uniformly thin-walled, lacking a pore, non-explosive and deliquescing to liberate ascospores within ascocarp; ascospores unicellular and frequently of bivalve construction; cleistothecia open by weathering or rupturing from internal pressure; conidia often present, phialidic in many. Includes 'green' and 'blue moulds'. 9 families.

ORDER ERYSIPHALES

Biotrophic parasites; ascocarps with 1 to several oval- to club-shaped explosive asci; ascospores unicellular, colourless; chains of conidia arising in basipetal succession from mother cell on superficial colourless mycelium; penetration of host by haustoria confined to epidermal cells. 'Powdery mildews'. 2 families.

Class Pyrenomycetes

Asci one-walled (unitunicate), cylindrical, explosive, opening by apical pore or slit, in hymenium with packing paraphyses in minute flask-shaped ascocarp (perithecium); perithecium opening by apical pore, dark- or bright-coloured, soft and fleshy or leathery, produced singly or in clusters, or in stroma; many with distinctive conidial states.

ORDER MELIOLALES

Tropical leaf parasites; dark mycelium forming superficial mat of thick-walled, regularly branched hyphae with hyphopodia arising from a short lateral cell; ascocarps superficial, dark, flattened or spherical; asci with thin evanescent walls, 2- or 4-spored; ascospores dark-brown, thick-walled, mostly 5-celled; conidial states absent. 'Black mildews'. 1 family.

ORDER CORONOPHORALES

Saprotrophic on wood; ascocarps dark carbonaceous, globose or turbinate, free or aggregated on compact stromatic structures, no true pore but opening by disintegration at apex induced by large gelatinous cells within swelling and forming mucilaginous mass; asci distinctly stalked, club-shaped and scattered throughout; ascospores variable, 1- to many-celled, colourless to pale-brown, cylindrical and curved, eventually filling ascus as powdery mass. 1 family.

ORDER SPHAERIALES

Perithecia flask-shaped, opening by apical pore, with bright fleshy or dark membranous or carbonaceous wall, solitary or aggregated or enclosed in stroma; asci and ascospores very variable in form. 15 families.

Class Discomycetes

Asci one-walled (unitunicate), cylindrical, explosive, operculate, opening by distinct lid or operculum, or inoperculate, opening by slit, arranged in hymenium with paraphyses, and exposed at maturity in cup- or saucer-shaped ascocarp (apothecium) (except in the Tuberales).

ORDER PEZIZALES

Asci operculate in distinct hymenium in apothecia of varying shapes, ranging from minute to large; saprotrophic on soil, dung, wood and plant debris. 10 families.

ORDER HELOTIALES

Asci inoperculate in distinct hymenium in apothecia of varying form; mostly saprotrophic but with a few plant pathogens. 8 families.

ORDER TUBERALES

With subterranean (hypogeal) ascocarps, modified apothecia, in which hymenium not exposed to exterior; asci non-explosive, without operculum, breaking down within ascocarp; asci 8-spored (or fewer), ascospore number often varying among asci within same ascocarp; ascocarps with strong smell and flavour, spore dispersal by rodents. 'Truffles'. 3 families.

ORDER PHACIDIALES

Plant pathogens; ascocarp begins as stroma, black, immersed wholly or partially in host tissue; hymenium exposed by cracks in surface of fungal stroma into 1 or more slits, or in stellate manner; asci thickened apically; ascospores often with gelatinous sheath. 3 families.

ORDER OSTROPALES

Ascocarps superficial or partially immersed in wood or leaves; differ from the Helotiales in possession of cylindrical, very long, narrow asci with strongly thickened apex and filiform asci nearly as long as ascus. 1 family.

ORDER CYTTARIALES

Parasitic on *Nothofagus* trees in the Southern Hemisphere; aggregates of ascocarps spherical or pear-shaped on galls of host tissue; apothecia, up to 200 in each, develop beneath cortex, each eventually opening separately as broad disc. 1 family.

ORDER LECANORALES

A very large group (about 29 per cent of all known species of fungi) of inoperculate mutualistic symbionts with algae in 'lichens'. The majority of lichenized fungi belong to this order. (Other fungi are also known to form lichens, e.g., some members of the Hysteriales, Pleosporales, Sphaeriales and Basidiomycotina.) 40 families.

Class Loculoascomycetes

Asci 2-walled (bitunicate), outer wall thin and inextensible, inner wall thick and extensible, seen as outer wall ruptures at apex and inner wall extends just prior to ascospore discharge; ascospores released through pore at tip of inner wall; ascocarp an aggregate of vegetative hyphae within which 1 or more locules develop by downgrowth of pseudoparaphyses attached at top and bottom (unlike true paraphyses which arise from basal hyphae and are free at tip); when single loculus develops, containing a number of asci, whole resembles true perithecium and is called a pseudothecium; ascospores septate and usually coloured.

ORDER MICROTHYRIALES

Mostly tropical or subtropical, on living leaves and stems; pseudothecia superficial, flattened, shield-shaped; asci globose to cylindrical; ascospores mostly uniseptate. 11 families.

ORDER MYRIANGIALES

Mostly tropical or subtropical; epiphytes, parasites or hyperparasites on superficial fungi or scale insects on living leaves and stems; asci globose, scattered individually throughout ascocarp. 4 families.

ORDER DOTHIDEALES

Asci ovoid, club-shaped or cylindrical, grouped in small locules without pseudoparaphyses in pseudothecia; pseudothecia separate or grouped on or in stroma; ascospores usually uniseptate. 8 families.

ORDER PLEOSPORALES

Asci long, cylindrical, separated by pseudoparaphyses in relatively large, uniloculate, usually solitary pseudothecia; ascospores commonly phragmosporous or dictyosporous, pigmented. 8 families.

ORDER HYSTERIALES

On dead, woody branches and bare wood; with distinct boat-shaped, carbonaceous pseudothecia opening by longitudinal slit and appearing apothecium-like when moist. 6 families.

Class Laboulbeniomycetes

Predominantly obligate parasites of hexapods, especially coleopterans with distinctive non-mycelial and determinate growth pattern. With main body of the fungus, the receptacle, attached to host by basal cellular holdfast; single, simple haustorium penetrating host; receptacle varies in size and complexity, in some row of 3 cells, in others large number of cells superimposed in tiers. Lateral filamentous appendages and 1 or more sessile or stalked perithecia arise on receptacle; asci usually 4-spored; ascospores commonly colourless, elongated and more or less spindle-shaped, 2-celled with large basal cell, each surrounded by colourless envelope thickened at lower end; ascus wall deliquesces prior to spore discharge.

A single order (Laboulbeniales), 4 families.

Phylum (Subdivision) Deuteromycotina

The so-called 'Fungi Imperfecti': fungi known only from their asexual (imperfect or anamorphic) or mycelial state. Their sexual (perfect or teleomorphic) states are either unknown or may possibly be lacking altogether. Solely conidial or mycelial fungi. Unrelated fungi from the Ascomycotina, Basidiomycotina and even the Zygomycotina may produce conidial states which are morphologically very similar and may be classified together in the same genus. Because any such genus may include unrelated taxa, it is called a 'form-genus'.

The 15 000 species are apportioned between 3 classes.

Class Blastomycetes

Yeast-like budding forms, true mycelium lacking or not well developed. This group contains not only asporogenous yeasts (those which do not form asci) but also yeast-like organisms, such as *Sporobolomyces*, which are clearly related to the Basidiomycotina in that they produce ballistospores, some have clamp connections and their DNA ratios are similar.

Class Hyphomycetes

Mycelial forms, with conidia borne directly on hyphae or on special hyphal branches, conidiophores; latter borne singly, aggregated together or in compact cushions.

Attempts have been made to produce a 'natural' classification but traditional hierarchical divisions into orders and families are usually abandoned. Form-genera are characterized by using four sets of criteria: (1) Saccardoan spore group based on septation and shape; (2) general arrangement of the conidia; (3) colour of conidia; and (4) type of conidial development. Conidia may be 1, 2 or many celled and may be divided by septa in 1–3 planes. They may be globose, ovoid, ellipsoid, cylindrical, branched or variously coiled. They may be arranged in a variety of types of chains or in slimy heads, or not in chains or slime. They may be colourless, brightly coloured or dark. Conidial development may be thallic, that is, where there is no enlargement of the recognizable initial, or, if there is, enlargement takes place after the initial has been delimited by a septum or septa; or it may be blastic where marked enlargement of the conidial initial takes place before it is delimited by a septum.

Class Coelomycetes

Mycelial forms, with conidia borne in flask-shaped pycnidia or saucer-shaped acervuli; pycnidia superficial or immersed, spherical, flattened or discoid, yellow, orange, brown or black, separate or aggregated, usually with single opening at apex; acervuli usually immersed, separated or aggregated together on stroma, variously coloured. The conidia show a similar range of shape, septation, colour and modes of development as those of the Hyphomycetes.

ORDER MELANCONIALES

Conidia produced in acervuli. 1 family.

ORDER SPHAEROPSIDALES

Conidia produced in pycnidia. 4 families.

Phylum (Subdivision) Basidiomycotina

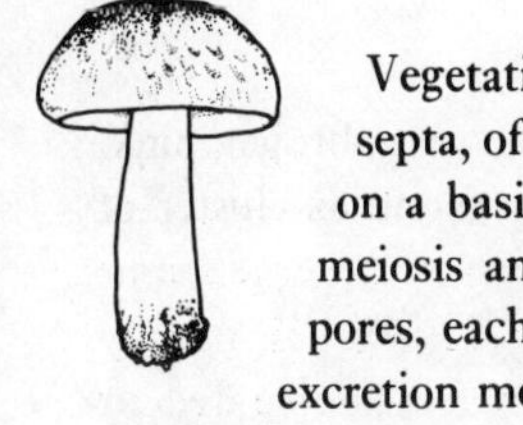

Vegetative dikaryophase; hyphae septate, with dolipore septa, often with clamp connections; basidiospores produced on a basidium; nuclear fusion in the basidium followed by meiosis and production exogenously typically of 4 basidiospores, each on a stalk, usually violently discharged by a drop-excretion method as ballistospores; wall of chitin and glucan. The 12 000 species are apportioned between 3 classes and 19 orders.

Class Hymenomycetes

Basidia exposed at maturity, arranged in palisade-like hymenium; basidiospores typically ballistospores, asymmetrically poised on their stalks. 'Mushrooms', 'toadstools', 'bracket fungi' or 'polypores', 'fairy clubs' and 'jelly fungi'. 2 subclasses.

Subclass Holobasidiomycetidae

Basidium a holobasidium: single, cylindrical cell, undivided by septa, typically bearing 4 spores on stalks at apex.

ORDER EXOBASIDIALES

Strictly plant pathogens; without well-defined basidiocarps; basidia occur in isolation, in fascicles or forming continuous hymenium, emerging through stomata or between epidermal cells. 1 family.

ORDER BRACHYBASIDIALES

Parasitic, causing leaf spots; minute pustulate to discoid, more or less gelatinous basidiocarps emerging through stomata or bursting through epidermis. 1 family.

ORDER DACRYMYCETALES

Saprotrophic on wood; basidiocarps gelatinous or waxy, often coloured yellow to orange; basidia forked or cleft of tuning-fork type, with 2 (often 3) septate, basidiospores. 1 family.

ORDER TULASNELLALES

Mostly saprotrophic; basidiocarps effuse, waxy, white, grey, brown, pink, violet or lilac; hymenium smooth, discontinuous or found as cluster of basidia. 2 families.

ORDER AGARICALES

Basidiocarps stalked like an umbrella, fleshy, composed of thin-walled, inflated hyphae (monomitic); hymenium usually lining gills, sometimes tubes, less frequently spines. 'Mushrooms' and 'toadstools'; 'agarics' and 'boletics'. 16 families.

ORDER APHYLLOPHORALES

Basidiocarps membranous, leathery, corky or woody, composed of thin-walled generative hyphae giving rise to thick-walled, unbranched, aseptate, skeletal hyphae or thick-walled, narrow, rarely septate, much-branched, binding hyphae or both (monomitic, dimitic or trimitic); hymenium usually lining tubes opening to exterior by pores, sometimes spines, anastomosing gills, folds or smooth surfaces, or erect branched or unbranched cylindrical or clavate basidiocarps. 'Polypores' or 'bracket fungi' and 'fairy clubs'. 23 families.

Subclass Phragmobasidiomycetidae

Basidiocarps gelatinous or waxy; basidium a phragmobasidium, divided by septa; basidiospores often germinate by repetition, i.e., by producing secondary spores. The Dacrymycetales are often included in this subclass.

ORDER TREMELLALES

Basidiocarps gelatinous, waxy or fleshy, often brightly coloured, drying to cartilaginous texture; basidia cruciately septate, i.e., usually longitudinally divided into 4. 3 families.

ORDER AURICULARIALES

Basidiocarps gelatinous to rubbery, drying to cartilaginous texture; basidia more or less cylindrical, with 1–3 (usually 3) transverse septa. 1 family.

ORDER SEPTOBASIDIALES

Mutualistic symbionts with scale insects; basidiocarps crustose or spongy; basidia more or less cylindrical, with 1–3 (usually 3) transverse septa. 1 family.

Class Gasteromycetes

Basidia not exposed at maturity, enclosed within cavities in closed basidiocarps of wide range of form and structure; basidiospores not violently discharged, liberated by collapse of basidium, usually symmetrically poised on their stalks or sessile, variously dispersed as dry powder, in slime, or within distinct packets (peridiola). 'Puffballs', 'earth stars', 'stinkhorns', 'bird's nest fungi', etc.

ORDER PODAXALES

Basidiocarps agaricoid in shape (*Coprinus*-like); gleba (spore mass) powdery at maturity, intermixed with tangled mass of sterile hyphae (capillitium). 1 family.

ORDER PHALLALES

Basidiocarp initially more or less globose, egg-like and buried (hypogeal); wall (peridium) of outer and inner papery layers with gelatinous material between; central part with compressed spongy or tubular tissue (receptacle) in addition to gleba; when peridium ruptures, receptacle expands carrying gleba out of 'egg'; gleba reduced to dark, often putrid-smelling slime, in which spores suspended. 'Stinkhorns'. 6 families.

ORDER HYMENOGASTRALES

Basidiocarps hypogeal, globose, up to 1.5 cm in diameters; peridium single or double; gleba fleshy to cartilaginous, sometimes powdery. 'False truffles'. 3 families.

ORDER LYCOPERDALES

Mature basidiocarps epigeal, earlier stages often hypogeal, usually sessile; peridium 2-layered; outer layer soon shed (puffballs) or splits stellately (earth stars), inner layer ruptures apically to form pore allowing spore release; gleba powdery at maturity with capillitium threads; basidiospores stalked and smooth-walled. 'Puffballs' and 'earth stars'. 4 families.

ORDER SCLERODERMATALES

Basidiocarps more or less globose; peridium 1-layered; gleba powdery at maturity, without capillitium; basidiospores sessile, with reticulate surface. 'Earthballs.' 4 families.

ORDER TULOSTOMATALES

Basidiocarps with more or less globose fertile part supported on well-developed stalk; peridium 1- or several-layered; gleba powdery at maturity, with capillitium. 'Stalked puffballs'. 2 families.

ORDER MELANOGASTRALES

Basidiocarps hypogeal; peridium 1- or 2-layered; gleba fleshy at maturity, series of cavities or islands with spore-bearing and non-spore-bearing basidia, latter may become converted to 'conidia'. 2 families.

ORDER NIDULARIALES

Basidiocarps globose or funnel-shaped, up to 1 cm in diameter, gregarious; peridium with 1 to many layers, dehiscing irregularly, by stellate apical splitting or by circumscissile epiphragm; remains of peridium forming cup ('nest') containing peridiola ('eggs'). 'Bird's nest fungi'. 2 families.

Class Teliomycetes

Mycelium septate with simple pore; equivalent of basidia consist of thick-walled teliospores or chlamydospores within which nuclear fusion takes place, and promycelia, which usually give rise to 4 or more spores after a meiosis; basidiocarps lacking. 'Rusts and smuts'.

ORDER UREDINALES

Biotrophic parasites of tracheophytes; with reddish-brown spore masses (urediospores); promycelium, from teliospore, tri-septate, producing typically 4 ballistospores; many with complex life-cycles involving 5 spore stages and 2 hosts. 'Rusts'. 5 families.

ORDER USTILAGINALES

Parasitic on angiosperms; with dusty, black spore masses (chlamydospores); promycelium, from chlamydospore, producing large, indefinite number of spores, often by budding. 'Smuts'. 2 families.

KINGDOM PLANTAE

R. S. K. BARNES

In the five-Kingdom system, the Kingdom Plantae corresponds to the 'higher green plants' (= archegoniates = embryophytes = cormophytes) of the two-Kingdom, plants-versus-animals, classification. Plants are, therefore, small to very large, multicellular, oxigenically photosynthetic, non-motile, eukaryote organisms with cellulose cell walls, a shoot usually differentiated into stem and leaves, and often with a root system, with (at least ancestrally) multicellular sporangia and gametangia enclosed within envelopes of sterile cells, and with a regular alternation of haploid and diploid generations. Haploid plants are gametophytes; diploid plants are sporophytes: the sporophyte generation meiotically produces haploid spores which, when they germinate, produce the haploid and often thallose gametophytes; these produce haploid gametes mitotically, which, when they pair and fuse, form diploid zygotes which develop, via embryos, into the sporophyte generation. In some plants, the two generations are completely separate organisms; in many, however, one generation is small to minute, and is attached to, and entirely dependent on, the other.

The Plantae is probably a monophyletic group descended from chlorophyte protists, with the bryophytes being descended from the same ancestral stock as the psilopsidan tracheophytes. The essence of their distinctiveness and success is that they are the photosynthetic organisms which have colonized the land: they are fundamentally terrestrial organisms, anchored in a substratum, and most of their structure relates to this. Like their chlorophyte ancestors, plants are obligately photoautotrophic and contain chlorophylls *a* and *b* (and other more minor pigments) in intracellular membrane-bounded chloroplasts. However, characteristically they also: (a) develop some form of supporting tissues involving a hardening of the cellulose cell walls; (b) develop a system of conducting tissues to transport substances from the sites of uptake or synthesis to the other tissues; (c) possess a surface resistant to water loss; and (d) have pores (stomata) through the epidermis to facilitate the exchange of gases. Many plants have secondarily returned to aquatic habitats in the form of freshwaters, but very few have recolonized the sea, and then only around its margins.

Phylum Bryophyta

Bryophytes are small plants in which the haploid gametophyte is the larger, more dominant generation. The small sporophyte is partially or totally dependent on the leafy or thallose gametophyte, on which it grows, absorbing nourishment through a foot and otherwise comprising a usually stalked, spore-forming capsule. The sporophyte, unlike the gametophyte, is never branched or leafy; it grows from the gametophytic archegonium and produces large numbers of spores in tetrads. The spores, either directly or via protonemata, develop into the mono- or dioecious gametophyte generation. Gametophytes lack true roots but possess simple or branched, hair-like anchoring rhizoids; lack xylem or phloem tissues but may have simple conducting bundles; may possess stomata; and multiply asexually by fragmentation and by the production of propagules (gemmae). Their gametangia are multicellular and are enclosed in an envelope of sterile cells: the antheridia are spherical to club-shaped, are stalked, and produce biflagellate sperm; the archegonia are flask-shaped and produce a single egg plus several canal cells. Fertilization requires the presence of external water and occurs within the archegonia; the zygote develops without any resting phase into an embryo, which grows into the sporophyte, eventually bursting through the archegonial wall.

Although terrestrial, bryophytes require moist surroundings for active life; a cuticle is present, but it is thin and delicate. Some can persist through dry periods in an inactive state, and can take up moisture when available again (from dew, from the atmosphere, etc.) over their whole surface. Others are restricted to permanently humid habitats; some are freshwater species.

The 23 000 species can be assigned to 3 classes.

Class Hepaticopsida

With thallose or leafy gametophyte attached to substratum by unicellular, unbranched rhizoids; without stomata; with oil bodies (spherical droplets of ethereal oils) in some or all cells; with many small, pyrenoid-less chloroplasts per cell; sporophyte short-lived, usually without chlorophyll when mature; capsule without stomata usually dehiscing along 4 lines, with sporogenous cells producing spores and sterile elaters; gametangia develop from superficial cells; protonemata usually small. 'Liverworts'.

ORDER JUNGERMANNIALES

Leafy, usually with 2 lateral rows and 1 ventral row (often reduced) of leaflets; stem without central conducting strand; antheridia in leaf axils; archegonia terminal, surrounded by perichaetium or perianth; capsule 4-valved; without stomata. 43 families.

ORDER METZGERIALES

Leafy or thallose; with flat axis and lateral wings, sometimes with ventral scales; archegonia non-terminal, dorsal, with base enclosed by involucral outgrowth of thallus; leaves not bilobed; stem sometimes with central conducting strand. 12 families.

ORDER CALOBRYALES

With prostrate basal leafless stolons and erect leafy branches; without rhizoids; mucilagenous; archegonia and sporophyte stalk massive; oil bodies minute or absent; dioecious. 2 families.

ORDER MARCHANTIALES

Large, thallose; thallus with dorsal epidermis and compact ventral tissue; epidermis with pores, photosynthetic tissue with layers of air chambers; ventral tissue with scales and dimorphic rhizoids; sporophyte with short or no stalk; archegonium usually elevated on branches of thallus. 12 families.

ORDER MONOCLEALES

Thallose; thallus dichotomously branching, solid, without epidermal pores, air chambers or midrib; tissue undifferentiated, without ventral scales; with dimorphic rhizoids; archegonia massive, with cylindrical involucra; sporophytes with long, massive stalks; capsule with 1 slit. 1 family.

ORDER SPHAEROCARPALES

Thallus simple, small, forming rosette or winged axis; without air chambers; with simple rhizoids; archegonia non-terminal, within involucra; sporophyte capsule not dehiscent; without elaters. 2 families.

Class Anthocerotopsida

Gametophyte leafless, prostrate, thallose, with undifferentiated tissue, ventrally with stomata, each with 2 guard cells; stomata lead into mucilage-filled cavities, cells each with 1 large chloroplast with pyrenoid, without oil bodies, ventral surface with simple rhizoids, without scales. Sporophytes slender, with persistent meristematic zones, indeterminate growth, chloroplasts and stomata; capsule dehisces into 2 long linear valves; sporogenous cells producing spores and elaters; gametangia develop from internal cells, sited dorsally; spore maturation not simultaneous. 'Hornworts'.

A single order (Anthocerotales), 1 family.

Class Muscopsida

Gametophyte leafy, radially organized, differentiated into stem and leaves, attached to substratum by multicellular, branched, filamentous rhizoids with oblique cross walls; develops via thallose or highly branched, green protonemata; with gametangia in groups on tips of main or lateral axes, supported by upper leaves forming a perianth, leaves spirally arranged. Sporophytes producing only spores (no elaters), capsule with stomata, each with 2 guard cells, on lower region, with active photosynthetic tissue, with central columella formed by endothecial cells; spermatozoids and eggs with a few, minute chloroplasts; with marked regenerative abilities. 'Mosses'.

In recent years, the mosses have been split into more and more, smaller orders: here they will be divided between 6 superorders, which perhaps correspond more with the nature of orders in other groups of organisms.

SUPERORDER SPHAGNANAE

Spores germinate in presence of mycorrhizal fungus to form thallose protonemata, each protonema giving rise to 1 gametophyte; rhizoids on thallus only; stem with core of cortical cells holding water; capsule elevated on gametophytic tissue, with massive columella, operculate, without peristome; antheridia stalked, in leaf axils; calcifuge; aquatic. 'Bog mosses'. 1 family.

Superorder Andreaeanae

Thallose protonemata branched, ribbon-shaped; gametophyte small, brittle, epilithic on non-calcareous rocks; capsule elevated on gametophytic tissue, dehisces along 4(–10) longitudinal lines, valves remaining attached above and below, with columella capped by bell-shaped spore sac. 2 families.

Superorder Tetraphidanae

Protonemata thallose; capsule elevated on sporophytic stalk which elongates before capsule develops, with peristome of 4 pyramidal teeth of numerous tiers of whole cells; with poorly developed gametophytic conducting tissue; without leaf lamellae. 2 families.

Superorder Polytrichanae

Protonemata filamentous; gametophyte with prostrate underground portion and aerial stems, well-developed conducting and supporting tissues, complex leaf structure and numerous lamellae; peristome with 16–64 teeth composed of bundles of whole cells from 4 layers of amphithecial tissue. 2 families.

Superorder Diplolepanae

Protonemata filamentous; gametophytic conducting tissue not well-developed, without leaf lamellae; peristome formed from 3 layers of amphithecial tissue, comprising, when not reduced, 2 concentric rows of teeth (outer with 2 times number of cell plates on outer surface as on inner). 64 families.

Superorder Haplolepanae

Protonemata filamentous; gametophytic conducting tissue not well-developed, without leaf lamellae; peristome formed from 2 layers of amphithecial tissue, comprising, when not reduced, 16 teeth. 21 families.

Phylum Tracheophyta

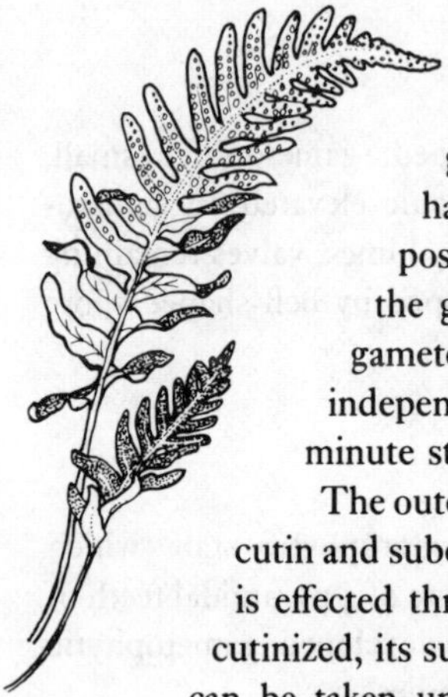

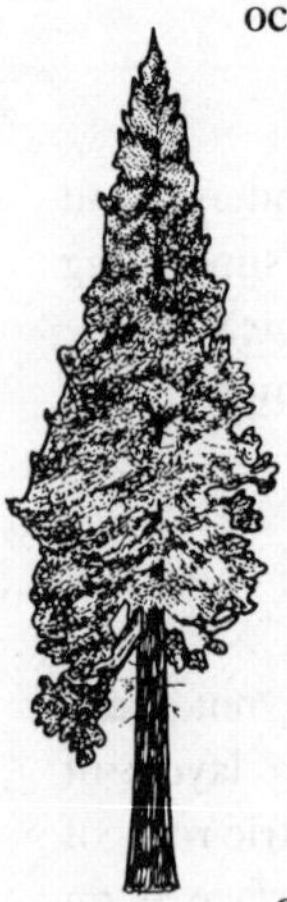

Tracheophytes are plants in which the diploid sporophyte is the larger more dominant generation: the sporophyte is small to very large, has a shoot divided into stem and leaves, and possesses an absorptive and anchoring root system; the gametophyte and gametangia are reduced, the gametophyte, in some, to a small prothallus living independently of the sporophyte, and, in others, to a minute structure enclosed within sporophytic tissues. The outer layer of sporophytic cells are waterproofed by cutin and suberin, thickened and cutinized; gaseous exchange is effected through closable stomata; the root system is not cutinized, its surface area is increased by root hairs, and water can be taken up across its surface; a water-transport system occurs in the form of tracheids, and organic materials are conducted in sieve cells or tubes; when aggregated into vascular bundles, tracheids comprise xylem tissue and sieve tubes phloem; support is achieved by the deposition of lignin in cell walls; vegetative multiplication of the sporophyte by fragmentation and by propagules is common, some species multiplying only asexually. The free-living gametophyte often contains chlorophyll, is filamentous or thallose, and is attached to the substratum by simple rhizoids; various stages of reduction occur, down to the male gametophyte being within a pollen grain or tube and the female gametophyte within an embryo sac. Polyploidy, especially allopolyploidy, is common.

The 10 classes of the monophyletic tracheophytes contain over 250 000 species and to some extent comprise a sequence of increasing morphological complexity and adaptation to terrestrial conditions (comparable to that seen in the tetrapod vertebrate animals). Of the 10 classes, those from the Psilotopsida to the Filicopsida (inclusive) in the list below share a number of common features and are often grouped together as the Pteridophyta, as distinct from the seed-producing Spermatophyta (classes Coniferopsida to Liliopsida, inclusive); within the spermatophytes, classes Coniferopsida to Gnetopsida (inclusive) have non-enclosed ovules and are sometimes united to form the Gymnospermae, whereas in the classes Magnoliopsida and Liliopsida (the Angiospermae) the ovules are enclosed within the carpels.

Class Psilotopsida

Prothalli small, cylindrical, branched; without chlorophyll; living underground with mycorrhizal fungus. Sporophyte herbaceous, dichotomously branching; without true roots; with filamentous mycorrhizal rhizoids; with vertically-flattened or minute scale-like nerveless leaves; sporangia thick-walled, united in twos or threes to form synangia on short stalks near forks of bifid leaves; homosporous. External water necessary for fertilization; polyploid; archegonia with 1–2 neck canal cells; prothalli large, with vascular bundles and lignified tracheids.

A single order (Psilotales), 1 family.

Class Lycopsida

Prothalli small, subterranean, saprophytic, or gametophytic generation passed within the sporangia; sporophyte with roots, stem and needle- or scale-like, spirally arranged leaves, each with 1 unbranched vein; stem without long internodes; sporangia in leaf axils or within leaf bases; sporophytes small.

ORDER LYCOPODIALES

Prothalli long-lived, mycorrhizal, tuber-like, monoecious, with numerous neck canal cells in archegonia; sporophytes evergreen, without secondary thickening, with needle- or scale-like leaves and adventitious roots on the creeping stem; sporangium-bearing leaves often forming apical cones; homosporous; terrestrial, some epiphytic. ‘Clubmosses’. 1 family.

ORDER ISOETALES

Prothalli dioecious, develop within spores; sporophyte with short stem, dense rosette of subulate leaves, each with dilated base, 4 air passages, ligule and, in outer ones, 1 sporangium imbedded in base; with 1 vascular bundle, cambium producing secondary thickening; roots branch dichotomously; heterosporous; aquatic or semi-aquatic. ‘Quillworts’. 1 family.

ORDER SELAGINELLALES

Prothalli dioecious, develop within spores; sporophyte with creeping stem, distal clusters of adventitious roots on rhizophores, scale-like leaves in 4 rows or spiral (sometimes of 2 types), with ligule; sporangia axillary in terminal cones; heterosporous; terrestrial. 'Spike mosses'. 1 family.

Class Sphenopsida

Small to medium sized; prothalli branched, lobular, green; sporophyte with perennial subterranean rhizomes from which issue erect, mostly annual stems with long internodes, whorls of pointed scale-leaves united at their bases at each node; stems grooved longitudinally, hollow, impregnated with silica; stems, not leaves, photosynthetic; sporangia on sporangiophores forming terminal cones, sporangiophores umbrella-shaped, with 5–10 sporangia beneath umbrella; homosporous, spores green, with haptera; spermatozoids with numerous flagella; terrestrial in moist habitats. 'Horsetails'.

A single order (Equisetales), 1 family.

Class Filicopsida

Small to large; prothalli free-living, usually photosynthetic, some colourless and subterranean, with gametangia on undersurface; sporophyte with creeping or erect stem, wiry or fleshy roots, with large fronds with many branching veins and usually a stalk; fronds with numerous sporangia on undersurface; fronds strap-shaped to multipinnate, arise from underground rhizome or massive trunk, mostly inrolled at tip (circinate) when young; usually homosporous; spermatozoids with numerous flagella. 'Ferns'.

ORDER OPHIOGLOSSALES

Prothalli subterranean, mycorrhizal, fleshy, monoecious; sporophyte with single annual frond with distinct fertile zone arising from base of lamina, sporangia massive and thick-walled, with subterranean, fleshy, mycorrhizal stems and fleshy roots; fronds not circinate. 1 family.

ORDER MARATTIALES

Small to very large; prothalli large, green, mycorrhizal; sporophyte with short tuberous stem, large fleshy pinnate fronds, large fleshy roots, with massive thick-walled sporangia in sori; sporangia of each sorus forming a synangium. 1 family.

ORDER FILICALES

Small to very large; prothalli usually green, heart-shaped; fronds with thin-walled sporangia opening by annulus; apogamy and apospory occur, particularly in polyploids. 23 families.

ORDER MARSILEALES

Heterosporous; prothalli within spores; sporophyte with long hairy rhizomes, fronds with long stalks and 0–4 leaflets at tip, with sporangia in hard-stalked sporocarps in pairs at base of leafy stalk; each sporocarp with many sori; each sorus with 1 megasporangium and many microsporangia; aquatic or semi-aquatic. 1 family.

ORDER SALVINIALES

Small; floating; heterosporous; sporophyte rhizome 'creeping' on water surface, roots or root-like leaves hang down in water, opposite or whorled fronds on or above water surface; sporangia in sporocarps, each sporocarp with 1 sorus; each sorus with either a few megasporangia or many microsporangia. 'Water ferns'. 2 families.

Class Coniferopsida

Medium to large sized; heterosporous; gametophyte enclosed within sporophytic tissue; sporophyte with well-developed roots, branched stem and leaves. Stem thickened by much secondary growth, xylem without vessels, resiniferous; leaves simple, scale- or needle-like, with 2 vascular bundles, often with heavy, waxy cuticle; microsporangia on microsporophyll scales, megasporangia (containing female gametophytes) on megasporophyll scales; scales aggregated on open, cone-shaped, micro- or megastrobili; wind pollinated, pollen tube growing through micropyle at base of exposed ovule (without ovary, style or stigma), fertilization often delayed, embryos enclosed in seed coat, each ovule with 1 megaspore; female gametophyte a multicellular prothallus with several archegonia; without spermatozoids, pollen tube conveys unflagellated sperm cells; free water not required for fertilization; embryo with 2 to many cotyledons. 'Conifers'.

ORDER PINALES

Stamens in dense spirals, numerous, dorsiventral, with pollen sacs on underside; megasporophyll scales forming cones, with ovuliferous scales in axils of bracts; ovuliferous scales with several ovules. 6 families.

ORDER TAXALES

Stamens peltate, each with 6–8 pollen sacs, with a few scale leaves at base; without female cones, ovules single, terminal, orthotropous, with a few scale leaves at base; seed surrounded by fleshy meristematic cup. 'Yews'. 1 family.

Class Ginkgoopsida

Large; heterosporous; with gametophyte enclosed within tissues of sporophyte; sporophyte with well-developed roots, branched stem, and leaves; stem thickened by much secondary growth, xylem without vessels, resiniferous; leaves fan-shaped, with parallel, dichotomously branching veins, in clusters on short spur shoots, deciduous; dioecious, paired ovules borne at apex of short stalk on spur shoot, males with similar short stalks with lax groups of microsporophyll stamens each with 2 pollen sacs; wind pollinated; without ovary, style or stigma, fertilization often delayed, embryos enclosed in seed coat, outer skin of ovule forming fleshy covering; gametophyte with many cells; spermatozoids large, with numerous flagella; free water not required for fertilization; 2 cotyledons. 'Maidenhair tree'.

A single order (Ginkgoales), 1 family.

Class Cycadopsida

Medium sized; heterosporous; gametophyte enclosed within tissues of sporophyte; sporophyte with well-developed roots, massive unbranched stem, and large, usually pinnate, initially circinate leaves in spiral near stem apex; stem with sparse vascular tissue, some secondary thickening, xylem without vessels, with pith and cortex forming most of stem; stem often short, sometimes subterranean, mucilaginous; large primary roots and small, secondary, coralloid roots on or above ground with symbiotic cyanobacteria; dioecious; microsporophyll stamens in spiral around elongate axis with sterile apex forming large, cone-like strobilus; ovules exposed on stalked scales on megasporophylls, 2 to many per sporophyll; sporophylls sometimes pinnate at apex, aggregated into strobili; usually wind pollinated; without ovary, style or stigma; pollen tube forms nourishment for male gametophyte, spermatozoids very large with numerous flagella, fertilization delayed; embryo enclosed in seed coat; female gametophyte massive, with several archegonia, each with very large egg; free water not required for fertilization; 2 cotyledons. 'Cycads'.

A single order (Cycadales), 3 families.

Class Gnetopsida

Heterosporous; gametophyte greatly reduced, enclosed within tissues of sporophyte; sporophyte with roots, stem and leaves of variable form; stem thickened by normal and adventitious secondary growth, secondary xylem with vessels, phloem sometimes with companion cells; dioecious (sometimes monoecious, rarely hermaphrodite); with perianth, male and female elements in compound, cone-like or spike-like inflorescences; male elements with 1–6 stamens usually with fused pollen sacs; female elements with single exposed ovule with 2 integuments and very long micropyle tube; ovule borne directly on floral axis; without ovary, style or stigma; micropyle secretes sugary droplets encouraging insect pollination; embryo enclosed in seed coat; no spermatozoids, pollen tube conveys unflagellated sperm cells; archegonia sometimes absent; free water not required for fertilization; 2 cotyledons.

ORDER GNETALES

Lianas, shrubs or trees; with opposite, elliptical, reticulate-veined leaves; inflorescences spike-like, in axils of united scale-leaves; flowers unisexual, male with 1 stamen; archegonia indiscernible. 1 family.

ORDER EPHEDRALES

Much-branched, green-shooted shrubs; with small, opposite or whorled scale-leaves; mostly dioecious, flowers solitary, paired or grouped, terminal or on branches arising from axils of bracts; male with 1 stamen; archegonia discernible. 1 family.

ORDER WELWITSCHIALES

With short stem, large tap root, 2 broad, continuously growing, strap-shaped, opposite leaves with parallel veins; flowers cone-like, in axils of bracts; male with 6 stamens and rudimentary ovule; archegonia indiscernible. 1 family.

Class Magnoliopsida

Small to very large; heterosporous; gametophyte greatly reduced, enclosed within tissues of sporophyte; sporophyte with well-developed roots, often branching stem, and leaves; stem with secondary growth, xylem with vessels in roots, stem and leaves, phloem with companion cells; roots primary and/or adventitious; leaves net-veined, usually with petiole, without basal sheath; reproductive organs aggregated into flowers, each flower basically with 2 perianth and 2 sporophyll whorls on tip of short shoot, whorls often of 4 or 5 units (or multiples)—whorl of green sepals forms calyx; whorl of (often) coloured petals forms corolla; whorl of microsporophyll stamens, each with 2 terminal pollen sacs, each pollen sac with 2 microsporangia, each pollen grain with 2 or 3 nuclei; whorl of megasporophyll carpels, each carpel comprising basal ovary, containing ovules, connected to distal stigma by slender style, each ovule with embryo sac, covered by 1–2 integuments; flowers often with nectaries (but not septal nectaries). Animal or wind pollinated; pollen grains usually with 3 germinal apertures, germinating on stigma; no spermatozoids, sperm cells without flagella; 2 nuclei enter ovule, one fusing with egg nucleus, other fusing with 2 embryo-sac nuclei to form triploid endosperm; without archegonia; free water not required for fertilization; embryo enclosed in seed coat and fruit derived from ovary; 2 cotyledons. 'Dicotyledons'.

In recent years, dicotyledons have been divided between an increasing number of smaller and smaller orders: here, they will be distributed between 16 superorders, which better correspond with the concept of orders in other groups of organisms.

SUPERORDER MAGNOLIANAE

Basal group, with free carpels, numerous spirally-arranged stamens, free petals, hermaphrodite flowers; pollen with 2 nuclei; stamens often not divided into filament and anther, perianth often not divided into calyx and corolla; woody; with ethereal oils; pollen grains with 1 germinal aperture. 27 families.

Superorder Nymphaeanae

Aquatic herbs, rooted in mud; without vessels or secondary growth; root hairs originate in specialized cells; free, spirally-arranged carpels; usually with marginal, superficial placentation; pollen grains with 1 or 3 germinal apertures; sometimes with trimerous flowers. 5 families.

Superorder Ranunculanae

Herbaceous; numerous stamens, isoquinoline alkaloids, free petals; pollen grains with 3 germinal apertures, pollen with 2 nuclei; usually with compound leaves. 10 families.

Superorder Hamamelidanae

Flowers reduced, often wind pollinated; woody tanniferous stem; with proanthocyanins and ellagic acid; flowers often without perianth, unisexual, forming catkin, ovary generally with 1 functional ovule. 24 families.

Superorder Rosanae

Carpels partly united; numerous ovules and centrifugally developing stamens; petals and sepals distinct. 31 families.

Superorder Myrtanae

Receptacles cup-shaped; numerous stamens; carpels fused; ovary inferior or semi-inferior, with central-axile placentation, ovules usually numerous; flowers actinomorphic, often tetramerous. 17 families.

Superorder Rutanae

Receptacle disc-shaped; perianth pentamerous; stamens in 2 whorls; ovary superior, with fused carpels; often with divided or compound leaves; sometimes with numerous ovules. 33 families.

Superorder Celastranae

Receptacle disc-shaped; perianth tetra- or pentamerous, actinomorphic; stamens usually in 1 whorl; few carpels and ovules; leaves simple, undivided; ovary superior or semi-inferior. 29 families.

SUPERORDER PROTEANAE

Ovary with 1 carpel, usually with 1 or 2 ovules; flowers tetramerous, without petals, bird or mammal pollinated; woody and tanniferous; with 3-celled trichomes. 1 family.

SUPERORDER ARALIANAE

Receptacle disc-shaped, with inferior ovary, 1 whorl of stamens; 1 pendulous ovule, often with 1 integument, in each loculus of ovary; flowers small, in umbels, with reduced calyx, usually hermaphrodite. 6 families.

SUPERORDER DILLENIANAE

Carpels usually fused, without septa; with separate petals; flowers regular, often with numerous stamens; ovules numerous, crassi- or tenuinucellar, with 1 or 2 integuments; leaves usually simple; often with mustard oils. 52 families.

SUPERORDER MALVANAE

Flowers pentamerous, actinomorphic, with separate petals; ovaries superior, with fused carpels; few carpels and ovules, ovules crassinucellar, with 2 integuments; frequently with mucilage cells. 4 families.

SUPERORDER ERICANAE

Flowers actinomorphic, with fused petals, 2 whorls of stamens, superior ovaries, often with 4 or 5 carpels; more stamens (incl. staminodes) than corolla lobes; leaves simple. 19 families.

SUPERORDER CARYOPHYLLANAE

Carpels fused, without septa, with central placenta; stamens often in 1 whorl, ovules reduced to 1 basal; petals separate; simple undivided leaves, betalains instead of anthocyanins, often with anomalous secondary growth; pollen with 3 nuclei. 14 families.

SUPERORDER LAMIANAE

Flowers actino- or zygomorphic, with fused petals, tetracyclic, with free anthers, fused carpels, few stamens, ovules with 1 integument, ovaries often inferior; stamens united with corolla; often with 2 carpels. 41 families.

SUPERORDER ASTERANAE

Flowers tetracyclic, with fused petals, with anthers fused postgenitally, filaments free; often with pseudanthia; ovaries inferior, often with 1 ovule, with brushes on style; stamens attached to corolla tube, equal in number or less and alternate with corolla lobes; often with inulin instead of starch, and latex canals. 9 families.

Class Liliopsida

Small to large; heterosporous; gametophyte greatly reduced, enclosed within tissues of sporophyte; sporophyte with well-developed roots, usually unbranched stem, and leaves; stem vascular bundles without cambium and without typical secondary growth, xylem often with vessels only in roots, never in leaves, phloem with companion cells; roots adventitious; leaves parallel-veined, usually without petiole, with basal sheath; reproductive organs aggregated into flowers, each flower basically with 2 perianth and 2 sporophyll whorls on tip of short shoot, whorls often of 3 units (or multiples)—whorl of green sepals forming calyx; whorl of (often) coloured petals forming corolla; whorl of microsporophyll stamens, each with 2 terminal pollen sacs, each pollen sac with 2 microsporangia, each pollen grain with 2 or 3 nuclei; whorl of megasporophyll carpels, each carpel comprising a basal ovary, containing ovules, connected to distal stigma by slender style, each ovule with embryo sac and covered by 1 or 2 integuments; flowers often with septal nectaries. Animal or wind pollinated; pollen grains usually with 1 germinal aperture, germinate on stigma; without spermatozoids, sperm cells without flagella, 2 nuclei enter ovule, one fusing with egg nucleus, other fusing with 2 embryo-sac nuclei to form triploid endosperm; without archegonia; free water not required for fertilization; embryo enclosed in seed coat and fruit formed from ovary; 1 cotyledon. 'Monocotyledons'.

In recent years, monocotyledons have been divided between an increasing number of smaller and smaller orders: here, they will be distributed between 6 superorders, the rank of which better corresponds with the concept of orders in other groups of organisms.

Superorder Alismatanae

Aquatic or semi-aquatic herbs; actinomorphic flowers, often unisexual or dioecious, with 2 + trimerous perianth whorls, often reduced; pollen with 3 nuclei; seeds usually without endosperm; vessels, if present, only in roots; leaves simple. 16 families.

Superorder Lilianae

All perianth elements petaloid; flowers actino- or zygomorphic, hermaphrodite, with septal nectaries; 3 variably-united carpels; with starchless (or without) endosperm; many species epiphytic or geophytic. 19 families.

Superorder Bromelianae

Perianth differentiated into petals and sepals; with septal nectaries, animal pollinated; pollen grains with 2 nuclei; 3 variably-united carpels in inferior ovary; seeds numerous, with starchy endosperm. 9 families.

Superorder Juncanae

Wind pollinated; embryo embedded in starchy endosperm; perianth of 2 trimerous whorls, inconspicuous, reduced; stems usually solid, without swollen nodes; leaves tristichous; ovary with 2 or 3 carpels; pollen in tetrads or monads. 2 families.

Superorder Commelinanae

Mostly wind pollinated; without nectaries; perianth often reduced and inconspicuous; leaves divided into sheath and lamina; stem often with swollen nodes; ovary superior, with 2 or 3 carpels; embryo lateral to starchy endosperm. 10 families.

Superorder Arecanae

Flowers inconspicuous on club-shaped spadix subtended by large spathe; ovaries superior, with 1 to few ovules, forming indehiscent fruits; endosperm, when present, with oils; leaves alternate, sometimes with petioles; with raphides; often wind pollinated. 7 families.

KINGDOM ANIMALIA

R. S. K. BARNES

In early classification schemes in which organisms were either 'animal' or 'plant', non- or unicellular, 'animals' formed the 'Protozoa' and the multicellular species then comprised the 'Metazoa'. The Kingdom Animalia of the five-Kingdom system includes only the metazoans.

Animals, therefore, may be defined as multicellular, obligately heterotrophic eukaryotes which develop, embryologically, by mitotic division of a zygote formed by the fusion of two gametes, one large (egg) and one small (sperm). Whilst many can also multiply asexually—by budding, fragmentation or parthenogenesis—all, at least primitively, are capable of sexual reproduction, the diploid zygote forming initially a hollow ball of cells, the blastula, and thereafter, by a diversity of embryological pathways, the multicellular adult. Members of this Kingdom are motile at some stage—and usually at all stages—of their life history, and most are bilaterally symmetrical, although some are without symmetry and others are radially or near-radially symmetrical.

The multicellular animal condition has probably arisen from within the ancestral Protista at least three times: the sponges (Porifera), coelenterates (Cnidaria and Ctenophora) and flatworms (Platyhelminthes) representing separate phylogenetic lines. The phyla Placozoa and Mesozoa may also have evolved independently from the protists. The other animal phyla (28 in this classification) have almost certainly been derived from a flatworm or flatworm-like ancestry. Attempts to group these diverse flatworm-derivatives into superphyla, mainly using embryological features, are many; such groupings as the 'acoelomates', 'pseudocoelomates', 'schizocoelomates', 'enterocoelomates', 'lophophorates', 'protostomes' or 'deuterostomes' have achieved wide usage. Opinions vary, however, on the affinities of many phyla, and as knowledge increases so many of the groupings are appearing more and more artificial; they are not used below. A broad picture of the phylogeny of the many phyla issuing from the flatworms is as follows (see, e.g., House, 1979; Barnes, 1980).

Largely, perhaps, as a response to predation pressure, some of the benthic flatworms evolved protective dorsal coverings (in effect becoming protomolluscs), and a wide spectrum of others evolved body cavities, which facilitated movement beneath the surface of the sediments, on the surface of which they had hitherto crawled. Different flatworm-like animals evolved hydrostatic skeletons by different embryological routes: today, some one dozen phyla of vermiform animals with fluid-filled body cavities are distinguished, *inter alia* on the nature of their cavities. These worms dominated the seas in the Cambrian and continue to dominate many marine areas at present.

Three large phylogenetic lines (or perhaps more correctly, three bundles of parallel lines) originated from this vermiform host in the Cambrian; all three associated with the development of hard protective shells or skeletons. One group of segmented worms underwent 'arthropodization', i.e., conversion of the mucopolysaccharide body covering into a chitinous and calcareous exoskeleton with the associated development of jointed limbs. This, like the evolution of body cavities, appears to have occurred across a broad front, although most lines with these adaptations became extinct before the end of the Devonian. A second series of lines led from a worm similar to the extant phoronids and gave rise to the various lophophorates encased in calcareous or phosphatic shells. The third line, characterized by various developmental peculiarities and by the tendency to produce subsurface calcareous plates, possibly also led from a phoronid-like worm and gave rise to the echinoderms, chordates and related groups.

The animals are the most diverse Kingdom of organisms, with over 100 groups of at least class rank, with more than 400 orders, and with several million species described and undescribed.

Phylum Placozoa

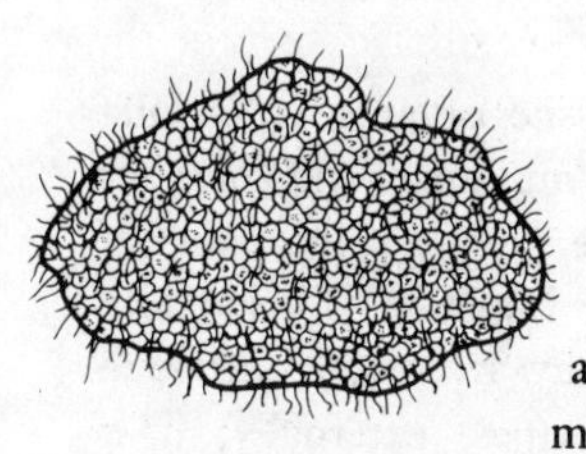

The phylum Placozoa was erected in 1971 for the single species *Trichoplax adhaerens* (and a doubtful second, *Treptoplax reptans*), known for a century but assumed to be the planuloid larva of an unknown metazoan. As yet known mainly from marine aquaria, in which it has been observed (and cultured) on many occasions, *Trichoplax* may either be the simplest animal known or, possibly, be related to the larval stage of a more complex species.

The animal is superficially similar to a giant amoeba (max. diameter 3 mm) but is composed of more than 1000 cells in the form of a flattened disc of variable outline. The flat upper surface is formed by a ciliated squamous epithelium; the somewhat concave lower surface by an epithelium mainly of ciliated columnar cells; between the two is a fluid-filled cavity with isolated stellate cells connected by mesenchyme. The body shape changes like that of an amoeba as the animal crawls in any direction. Feeding is by absorption of the products of extracorporeal digestion, enzymes being secreted into the space between the lower surface and the substratum. No organs or distinct tissues are present, neither are muscular or nervous systems. Both asexual fission and sexually-produced gametes are known.

Phylum Porifera

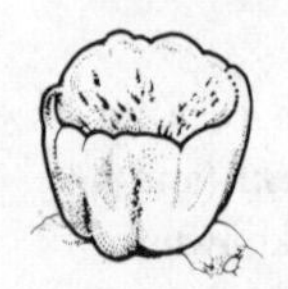

Poriferans, or 'sponges', are sessile organisms, lacking distinct tissues or organs, and muscular and nervous systems. Except in the larval stage, they are incapable of movement and indeed their whole life-style is related to immobility. The sponge body does not show any symmetry, although individual species may attain characteristic shapes; externally, it is perforated by a series of pores (through which water is induced to enter) and by one or more larger holes (through which the water leaves); internally, it takes the form of a series of canals and chambers through which water passes, drawn by the random beating of the flagella of choanocyte cells which line the canals or occur in some or all of the chambers. The integrity of the canals and body is maintained by an immovable skeletal matrix (of calcareous and/or siliceous spicules and/or organic fibres) which comprises the greater part of the sponge mass. Few types of cell are possessed; the choanocytes, which are almost identical to the free-living protistan choanoflagellates (p. 80), being the most complex. Most species are hermaphroditic. All are benthic filter feeders.

The 10 000 or so species (all but 150 freshwater species being marine) are apportioned between 4 classes and 26 orders.

Class Hexactinellida

Skeleton of siliceous spicules, mainly six-rayed, often fused together into a rigid, cup- or vase-shaped structure. Outer body surface without epidermis; covered by syncytium of interconnecting pseudopodia of amoebocytes. Choanocytes only in finger-shaped chambers. Spicules in two size-classes: large, six-rayed megascleres; and microscleres in the form of miniature megascleres (hexasters) or of single rods expanded into mushroom-like structures at each end (birotulates).

Entirely marine, mostly 500–1000 m but also in deep sea.

ORDER AMPHIDISCOSIDA

Megascleres not fused together; microscleres only as birotulates; never attached to hard substrata, always anchored in soft sediments by basal tuft(s) of spicules. 3 families.

ORDER HEXACTINOSIDA

Structural megascleres fused into rigid skeleton; microscleres only as hexasters; fusion of megascleres by cementation of adjacent rays lying side by side with secondary deposits of silica; always attached to hard substrata. 6 families.

ORDER LYCHNISCOSIDA

Structural megascleres fused into rigid skeleton; microscleres only as hexasters; fusion of megascleres occurs between characteristic lychniscs (the central part of each spicule is enclosed within a system of 12 struts arranged as along margins of an octohedron); always attached to hard substrata. 3 families.

ORDER LYSSACINOSIDA

Microscleres only as hexasters; megascleres usually free, sometimes secondarily and irregularly fused at base of body; some (sometimes all) structural megascleres monaxonic; attached to hard substrata or anchored in soft sediments. 4 families.

Class Calcarea

Spicules of calcite or aragonite, either free or secondarily fused, forming calcareous skeleton; spicules of variety of shapes but not divisible into mega- and microscleres. Epidermis of pinacocytes present.

Entirely marine, in shallow water (usually above 1000 m).

ORDER CLATHRINIDA

Choanocyte nucleus basal and independent of flagellum; choanocytes not confined to pockets but line internal canals. Larva flagellate over whole surface. Triradiate spicules, if present, with rays of equal length subtending equal basal angles. 1 family.

ORDER LEUCETTIDA

Choanocyte nucleus basal and independent of flagellum; choanocytes confined to chambers. Larva flagellate over whole surface. Triradiate spicules, if present, with rays of equal length subtending equal basal angles. 3 families.

ORDER LEUCOSOLENIIDA

Choanocyte nucleus apical, with flagellum arising from it; choanocytes not confined to pockets but line internal canals. Only anterior hemisphere of larva flagellate. Triradiate spicules with one ray longer than others and with unequal basal angles. 1 family.

ORDER SYCETTIDA

Choanocyte nucleus apical, with flagellum arising from it; choanocytes confined to chambers. Only anterior hemisphere of larva flagellate. Some triradiate spicules with one ray longer than others and with unequal basal angles. 5 families.

ORDER INOZOIDA

Choanocyte nucleus basal or apical. Larva flagellate over anterior hemisphere or all over. Skeleton with much calcareous material added as surface scales or plates, or as fused or reinforced tracts of spicules; spicules shaped like tuning forks in distinct tracts or randomly scattered. Confined to shaded caves in very shallow water. 5 families.

ORDER SPHINCTOZOIDA

Widely known from fossils, but with single living species (*Neocoelia crypta*). Body with segmental growth—linear series of chambers produced, solid aragonitic deposit filling oldest chambers. The living species shows similarities with the Demospongiae and, therefore, position of the order is uncertain.

Class Demospongiae

Skeleton of spongin fibres, sometimes reinforced with siliceous spicules in two size classes, mega- and microscleres; some genera without any skeletal material. Epidermis of pinacocytes present.

More than 90 per cent of sponges belong to this class which occurs in both the sea and freshwater.

ORDER HOMOSCLEROPHORIDA

Primitive demosponges without spongin fibres, with all spicules, if present, very small and often three-rayed; differentiation into mega- and microscleres not marked. Larvae incubated. 2 families.

ORDER CHORISTIDA

Coarse-textured, with megascleres showing some radial arrangement, if only peripherally and at the surface; without spongin fibres; megascleres four-rayed, with one long shaft and three short rays at one end, or monaxonal; microscleres star-shaped. Eggs extruded and develop externally. 7 families.

ORDER SPIROPHORIDA

Spherical, with radially arranged megascleres; without spongin fibres; megascleres four-rayed with one long shaft and three short rays at one end, or monaxonal; microscleres contorted sigmas covered by minute spines. Eggs extruded or retained within body. Typically on soft substrata anchored by basal mat of tangled spicules. 1 family.

ORDER LITHISTIDA

Polyphyletic group, probably derived from several demospongian orders, without spongin, with characteristic spicule, the desma, which has a complex pattern of branching; branches of adjacent desmas interlock or articulate forming a rigid skeleton. 9 families.

ORDER HADROMERIDA

With spongin but not as distinct fibres; megascleres in form of single rays pointed at one end and knobbed at other; megascleres arranged radially, often throughout body, sometimes only at surface; microscleres, if present, star-shaped. Eggs extruded. 10 families.

ORDER AXINELLIDA

Rough-textured, with a spongin and monaxonal spicular skeleton in form of a stiff axis and a plumose or plumo-reticulate extra-axial system issuing from it (extra-axial elements project through the body surface); microscleres often absent. 9 families.

ORDER AGELASIDA

With a reticulate, fibrous skeleton of spongin spined by unique monaxonal spicules pointed at one end, blunt at other and bearing whorls of small spines along their length. Eggs extruded. 1 family.

ORDER HALICHONDRIDA

With skeleton of spongin and monaxonal megascleres, no microscleres; except at surface, skeletal elements unorganized. Parenchymella larvae, ciliated over whole surface, incubated within body. 2 families.

ORDER POECILOSCLERIDA

With skeleton of spongin and monaxonal megascleres; megascleres regionally differentiated in form, often bearing surface spines; microscleres sigmoid, bow-shaped or chelate (with recurved ends). Parenchymella larvae, lacking cilia over posterior pole, incubated. 12 families.

ORDER PETROSIIDA

With reticulate skeleton of spongin and monaxonal megascleres, spicular element most important and largely confined to distinct tracts; monaxonal megascleres pointed at each end; microscleres, if present, bow-shaped. Larvae unknown. 1 family.

ORDER HAPLOSCLERIDA

With reticulate skeleton of triangular, rectangular or polygonal meshes formed of monaxonal megascleres joined at their ends by spongin; megascleres pointed at each end; microscleres, if present, sigmas or bow-shaped. Parenchymella larvae, with incomplete ciliation, incubated. Order includes the freshwater sponges (as well as marine species). 7 families.

ORDER VERONGIIDA

Biochemically and cytologically distinctive sponges of uncertain affinity. Spicules absent; spongin fibres often without outer layer. Eggs extruded. 3 families.

ORDER DICTYOCERATIDA

Without spicules but with complex anastomosing spongin-fibre skeleton; spongin fibres of at least two sizes. Large parenchymella larvae with very long posterior cilia, incubated. 3 families.

ORDER DENDROCERATIDA

Without spicules but with skeleton of spongin fibres arising from a basal plate and branching dendritically; fibres of single size. Large parenchymella larvae, with incomplete ciliation, incubated. 3 families.

Class Sclerospongiae

Essentially similar to the Demospongiae but with a basal skeleton of calcium carbonate. Living tissue, with skeleton of spongin fibres and siliceous spicules, penetrates only into upper layer of basal calcareous material; lower layers of calcium carbonate separated from living tissue.

ORDER CERATOPORELLIDA

With spongin fibres and spicules entrapped in aragonitic basal skeleton; living tissue a surface veneer. Individual units multiply by longitudinal fission. Spicules with spines in whorls or scattered along their length. 4 families.

ORDER TABULOSPONGIDA

With lamellar, calcitic basal skeleton; walls of skeletal units with spines; lower calcitic regions walled off by trabeculae. Individual units multiply by intramural budding. Free megascleres and microscleres similar to those of Hadromerida. 1 family.

Phylum Cnidaria

Cnidarians are animals with a radial or near-radial symmetry and, basically, with bodies in the form of a simple, elongate or flattened tube, closed at one end and open at the other, the margins of the open end being drawn out into a series of tentacles. The wall of this tube is divided into three zones: an outer epidermis; an inner layer of cells, the gastrodermis, lining the lumen of the tube; and, between the two, a layer of gelatinous material, the mesogloea, in which cells may or may not occur. The lumen of the tube, the coelenteron, forms the gut, into and out of which food passes through the mouth, the food being living animals which are caught by the tentacles circling the mouth. The agents of offence (and defence) are specialized cells, the cnidocytes, containing nematocysts (intracellular organelles, coiled in a cavity until fired, but on firing capable of being forcibly everted). Other specialized cells are nerve cells which form a simple nerve net, and epitheliomuscular cells which act against the mesogloea or water trapped in the coelenteron.

The cnidarian life-cycle involves an alternation of body form and reproductive type. In the asexual polyp phase, the tube-shaped body is attached to the substratum by its aboral end and the tentacle-fringed mouth is directed into the overlying water. Individual polyps may bud others to form a colony, sometimes with polymorphism of polyp type; or polyps may, by transverse fission or other asexual means, bud free-swimming forms (the medusae) in which the basic tube shape is flattened into a disc or bell, with a well-developed mesogloea, and a ventrally positioned mouth. The medusae bear gonads, and the zygotes develop, via a planula larva, into the attached polyp phase to complete the cycle. Different groups of cnidarians are characterized by dominance of either the polyp or the medusoid phase.

The 10 000 living species of cnidarians—all but a few of which are marine—can be divided into 2 subphyla, 6 classes and 29 orders.

Subphylum Medusozoa

Cnidarians in which the medusa is an important phase in the life-cycle, although it may secondarily be lost or reduced to a bud on the polyp.

Class Hydrozoa

Both polyp and medusoid phases occur, although either may be reduced. Mesogloea acellular; cnidocytes only in epidermis; gametes develop epidermally. Polyp phase: radially symmetrical, simple, with tubular coelenteron not partitioned by septa; often protected by chitinous calyx which may be calcified; typically colonial, with individual polyps connected by tubular extensions of the body; individual polyps specialized for different functions (defensive, reproductive, nutritive, etc.). Medusoid phase: small, transparent, usually with mouth on hollow stalk, with shelf of tissue (the velum) extending inwards from margin of bell.

ORDER ATHECATA

Polyps solitary or colonial, attached to hard substrata; stalks of polyps, but not distal tentacle-bearing portions, protected by chitinous calyx. Medusoid phase sometimes lacking, if present tall and bell-shaped; gonads on mouth stalk or stomach; without statocysts, with few peripheral tentacles. 34 families.

ORDER MILLEPORINA

Hydrozoans related to the Athecata and often included therein. Polyps colonial, embedded in large, calcareous matrix secreted within epidermis and covered by thin epidermal layer; individual polyps with small capitate tentacles on column as well as around mouth; inflict a powerful sting. Medusa small, without mouth, tentacles or velum. 'Fire-corals'. 1 family.

ORDER STYLASTERINA

Hydrozoans related to the Athecata and often included therein. Polyps colonial, embedded in large, calcareous matrix secreted within epidermis and covered by thick epidermal layer; individual polyps with or without tentacles, if present small and filiform. Medusa reduced to buds on specialized polyps. 1 family.

ORDER CHONDROPHORA

Hydrozoans related to the Athecata and sometimes included therein. Variously regarded as a single large polyp or as a colony of polyps, body a disc floating on sea surface, with or without an oblique sail, from disc hang numerous club-shaped tentacles and central mouth (or defensive and feeding polyps); gonads between tentacles and mouth. No medusoid phase. 2 families.

ORDER THECATA

Polyps colonial, attached to substratum; chitinous calyx extends as a cup or vase around distal parts of polyp. Medusa, if present, flattened, with gonads on radial coelenteric canals, with many tentacles and usually with statocysts. 22 families.

ORDER LIMNOMEDUSAE

Polyps with few or no tentacles, with thin or no calyx. Medusa with hollow tentacles, with internal statocysts. Mostly freshwater, but with some marine species. 4 families.

ORDER TRACHYLINA

A possibly artificial group of three types of hydrozoans—Laingiomedusae, Narcomedusae and Trachymedusae—in which medusa a modified actinula larva and not homologous with medusa of other hydrozoans. Polyp reduced or absent. 9 families.

ORDER SIPHONOPHORA

Colonial, polymorphic, with floating or swimming colonies containing, at the same time, feeding, stinging and reproductive polyps and swimming, buoyant and bract-like medusoid individuals, all budded from founding larva. 15 families.

ORDER ACTINULIDA

Very small (< 1.5 mm), solitary, with larval features (ciliated surface) and those of both polyps and medusae. Interstitial. 2 families.

Class Scyphozoa

Medusa dominant phase, polyp small and short-lived or absent. Medusa large and free-living (attached to substratum by aboral stalk or disc in one group), with gastrodermal gonads, cellular mesogloea and without velum. Polyp with coelenteron partitioned by 4 septa. 'Jellyfish'.

ORDER STAUROMEDUSAE

Medusa attached aborally to seaweeds or hard substrata, with 8 'arms' bearing small tentacles, usually with adhesive anchors between arms. High latitudes only. 2 families.

ORDER CORONATAE

Bell divided into upper and lower regions by groove; margin of bell deeply scalloped; tentacles solid, alternating with sense organs around margin. Mainly deep sea, often brightly coloured. 7 families.

ORDER SEMAEOSTOMEAE

Corners of single mouth drawn out into 4, broad, gelatinous, frilly lobes; hollow marginal tentacles evenly arranged around margin, in bunches or absent. 3 families.

ORDER RHIZOSTOMEAE

Mouth region with 4 long lobes which divide into 8 thick arms which fuse obliterating original mouth; many small openings in the arms form secondary, suctorial mouths; without tentacles. 10 families.

Class Cubozoa

Jellyfish intermediate between Scyphozoa and Hydrozoa. Medusa square in section, with 4 flat sides; unfrilled margin of bell drawn inwards to form velum-like organ; tentacle or group of tentacles at each corner of bell; nematocysts causing excruciating pain or death. Contains single order (Cubomedusae) with 2 families.

Subphylum Anthozoa

No medusa phase. Polyp large (sometimes very large) with thick, cellular mesogloea, and coelenteron partitioned by mesenteries bearing glands, filaments and cnidocytes; a sleeve-like pharynx or stomodaeum extending from mouth more than halfway down coelenteron, with 1 or 2 ciliated grooves (siphonoglyphs) directing water currents into coelenteron; an expanded zone around mouth, the oral disc, surrounded by tentacles; with epidermal and gastrodermal cnidocytes, differing from those of Medusozoa in lacking an operculum. Frequently colonial, but polymorphism less marked and less frequent than in Hydrozoa.

Class Ceriantipatharia

Musculature unspecialized; arrangement of mesenteries and tentacles simple; mesenteries single not paired, and complete (extend from stomodaeum to body wall); 6 primary mesenteries, if further ones added then they are added immediately opposite the single siphonoglyph.

ORDER ANTIPATHARIA

Erect whip-like or branching colonies attached to hard substrata, with black or brown, horny, thorn-bearing axial skeleton and small polyps with up to 24 non-retractile tentacles. 'Black' or 'Thorny corals'. 3 families.

ORDER CERIANTHARIA

Solitary anemone-like polyps inhabiting tubes in soft sediments, with larvae resembling antipatharian polyps. 2 rings of oral tentacles, inner ring of short tentacles and outer ring of more numerous long tentacles. No aboral pedal disc. 3 families.

Class Alcyonaria

Colonial, with internal skeleton of calcium carbonate or horny material. Polyps with 8 pinnate tentacles and coelenteron divided into 8 compartments by mesenteries, each mesentery with retractor muscle on side facing the single siphonoglyph. Polyp form varies little, alcyonarian classification being based on form of colony and its supporting skeleton.

ORDER PROTOALCYONARIA

Deep-water, solitary polyps of alcyonarian form. 2 families.

ORDER STOLONIFERA

Colonies with creeping stolons from which polyps arise separately; skeletal material, if present, calcareous spicules, free or fused into tubes (as in 'Organ-pipe coral'). 3 families.

ORDER TELESTACEA

Colonies with elongate axial polyps arising from creeping stolon and secondary polyps budded off laterally; skeleton of calcareous spicules joined together by horny or calcareous material. 3 familes.

ORDER GASTRAXONACEA

Colony with single axial polyp extending whole height and, distally, lateral polyps connected to, but not budded from, axial polyp. 1 species.

ORDER GORGONACEA

Colonies with a central stem attached to hard substrata from which arise branches, often in single plane. Stem with central stiffening rod of gorgonin, with or without calcareous spicules; short polyps on branches, not on main stem. 'Sea fans' and 'Sea whips'. 18 families.

ORDER COENOTHECALIA

Massive calcareous matrix penetrated by vertical, cylindrical cavities in which polyps situated. 'Blue coral'. 2 families.

ORDER ALCYONACEA

Colonies with lower portions of polyps forming rubbery or fleshy mass with calcareous spicules, into which distal portions can retract. 'Soft corals'. 6 families.

ORDER PENNATULACEA

Colonies with single large axial polyp (sometimes >1 m long) forming, basally, a stalk and terminal bulb for anchorage in soft sediments and, distally, bearing many short lateral polyps. Coelenteron of axial polyp with skeletal axes of calcified horny material in canals. 'Sea pens' and 'Sea pansies'. 14 families.

Class Zoantharia

Solitary or colonial anthozoans with secreted supporting material, if present, calcareous and external to the polyp body. Polyps with 6 (or multiples of 6) simple tentacles, with a coelenteron divided into numerous compartments by paired septa in multiples of 6; septa bearing retractor muscles such that those of each pair face each other; with 0–2 siphonoglyphs.

ORDER ACTINIARIA

Solitary polyps without secreted supportive material, with aboral end either as an adhesive disc for attachment to hard substrata or as a bulbous organ for penetrating soft sediments; typically with 2 siphonoglyphs. 'Sea anemones'. 41 families.

ORDER ZOANTHINARIA

Colonial or solitary sea-anemone-like polyps without secreted supportive material, without an aboral disc, and with 1 siphonoglyph. Colonial forms connected by basal stolon or mat. Attached to hard substrata or epizoic on benthic invertebrates. 3 families.

ORDER SCLERACTINIA

Solitary or, more commonly, colonial anthozoans with massive calcareous calyces including plate-like extensions inside the mesenteries; polyps occupy cups in calcareous matrix into which they can retract. No siphonoglyphs. 'Stony corals'. 23 families.

ORDER CORALLIMORPHARIA

Solitary or colonial polyps resembling scleractinians, but without calcareous material. Tentacles frequently with clubbed tips. 3 families.

ORDER PTYCHODACTIARIA

Poorly-known deep-sea, sea-anemone-like polyps without basilar muscles, without ciliary tracts on mesenterics, and with gonads arranged as in alcyonarians. 1 family.

Phylum Ctenophora

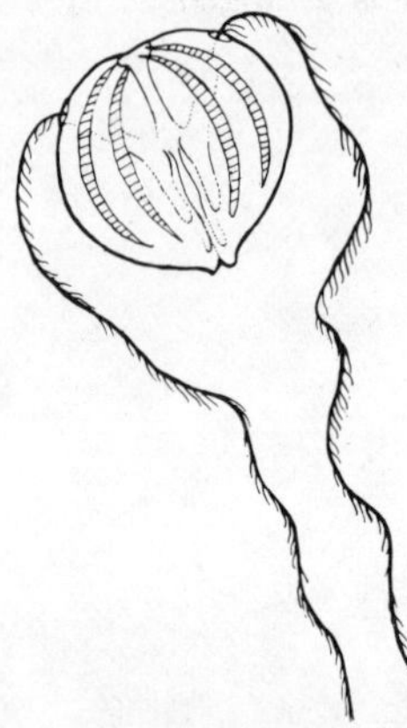

Predatory, pelagic marine animals with a bodily organization basically comparable to that of the cnidarian medusa, i.e., radial symmetry (in Ctenophora rendered biradial by the presence of 2 tentacles), two cell layers separated by a thick gelatinous mesogloea, and a nervous system in the form of a diffuse net. In ctenophores, however, the mesogloea contains muscle fibres and connective tissue as well as amoebocytes, the coelenteron possesses anal pores as well as a mouth, cnidocytes are not present (except in 1 sp.), and no polyp phase occurs.

The spherical to ribbon-shaped 'sea-gooseberries' or 'comb-jellies' possess eight rows of fused ciliary plates (comb plates) extending from aboral to oral pole. These beat causing the ctenophore to move mouth first, co-ordination being achieved via an apical statocyst. Typically, a pair of long retractile tentacles occurs, housed in tentacle sheaths. These branched structures contain lasso cells, the ctenophore-equivalent of cnidocytes, which when discharged produce a sticky thread; when extended, the tentacles are tens of times the body length. The gut possesses a series of blind-ending canals. Most ctenophores are simultaneous hermaphrodites, with external fertilization.

The 100 known species, all exclusively marine, are grouped in 7 orders.

Class Tentaculata

Possessing 2 tentacles.

ORDER CYDIPPIDA

Mostly spherical or oval-bodied, some flattened and slug-like, with long tentacles, well-developed tentacle sheaths, well-developed comb rows. (All tentaculate ctenophores pass through a larval stage with cydippid characteristics, and some cydippids may be the larval stages of members of other orders.) 5 families.

ORDER PLATYCTIDA

Bodies greatly compressed in the oral/aboral plane, with reduced or absent comb rows, with permanently or temporarily everted oral region forming a creeping sole. Asexual multiplication recorded; internal fertilization and brooding of larvae known. 4 families.

ORDER LOBATA

Bodies compressed in tentacular plane and expanded on each side of mouth into 2 oral lobes; as oral lobes develop, tentacles migrate orally, become reduced, lose their sheaths, and surround mouth. Ends of some gut canals anastomose. 6 families.

ORDER GANESHIDA

Little-known ctenophores with features of both cydippids and lobatans. Body compressed in tentacular plane; without oral lobes; during development tentacular sheaths reorganized and gut canals near mouth fuse. 1 family.

ORDER THALASSOCALYCIDA

Body expanded to form bell around mouth (food is trapped in mucus on inner surface of bell and passed by cilia to mouth), flattened in plane of stomach; tentacles near mouth, without sheaths. 1 species.

ORDER CESTIDA

Body greatly drawn out in tentacular plane forming ribbon up to 2 m long. 4 comb rows reduced, swimming achieved by muscular undulations of body, other 4 comb rows maintain orientation in water. Tentacles with sheaths present but much reduced; small accessory tentacles along lower (oral) body margin. 1 family.

Class Nuda

Without tentacles at any stage of life history. Body compressed somewhat in tentacular plane, with large stomodaeum occupying most of body volume and a wide and flexible mouth. Predators of other ctenophores. A single order (Beroida). 1 family.

Phylum Platyhelminthes

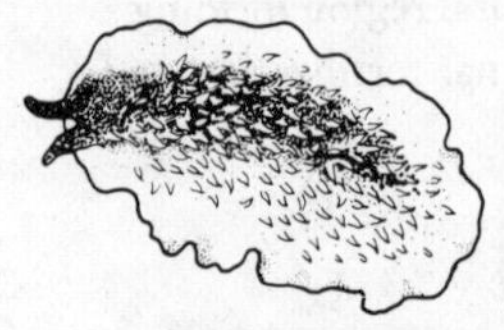

Platyhelminthes or flatworms are bilaterally symmetrical, soft bodied, vermiform animals with a dorso-ventrally flattened body. Organ systems occur, set in parenchymatous tissue of mesodermal origin—no body cavity occurs—and free-living, and some parasitic species, have a blind-ending gut. Except in some parasitic species, a distinct head occurs, bearing simple sense organs. All are feeders on animal tissue, whether as carnivores, scavengers or parasites. Free-living species are benthic and most are marine, although both freshwater and terrestrial species occur.

The gut, lacking in some groups, is sac-like and may be highly branched, the mouth is typically ventral; no circulatory or respiratory systems occur; excretion/osmoregulation is via protonephridia; and the relatively simple nervous system includes longitudinal nerves. Most species are hermaphroditic with complex sex organs and, in many, mechanisms of internal fertilization occur; the sperm are frequently biflagellate and cleavage of the egg is spiral. Parasitic species have a complex life-cycle, often involving several hosts, several larval stages and phases of asexual multiplication. Some parasitic species have lengths well in excess of 5 m; many free-living species are less than 1 mm long. All are capable of extensive regeneration of lost tissues and may consume their own tissues in the absence of food.

The 25 000 living species are distributed between 5 classes and 34 orders.

Class Turbellaria

Free-living flatworms with a ventral mouth, eversible pharynx, and, in most species, with a lobed or branched intestine; cellular and ciliated epidermis; with single (or bundles of) rhabdites—rod-like structures of uncertain function—in skin. Movement is effected by cilia of ventral surface and/or muscular contractions.

ORDER ACOELA

Small, often interstitial, marine; without permanent gut cavity, epidermal basement membrane, protonephridia, oviducts, yolk glands or membranes around gonads; statocyst present; bases of cilia interconnect. For many years considered to be syncytial because glycocalyx which takes up the stain absent; cell membranes have, however, been revealed by electron microscopy. 16 families.

ORDER CATENULIDA

Elongate, delicate, mainly freshwater; with simple pharynx, sac-like gut; without oviducts or yolk glands; with single protonephridium and unpaired gonads; male gonopore dorsal; sperm without flagella. 5 families.

ORDER MACROSTOMIDA

Marine and freshwater; with simple pharynx, sac-like gut, paired excretory ducts and complete reproductive system except yolk glands; without statocyst; penis with stylet; sperm without flagella. 3 families.

ORDER HAPLOPHARYNGIDA

Interstitial; similar to macrostomids but with a simple proboscis opening just beneath the anterior tip of body, and a permanent but poorly differentiated anal pore. 1 family.

ORDER ARCHOOPHORA

Small, marine; with elongate, plicate pharynx, sac-like gut, large frontal gland and simple male reproductive system opening posteriorly; without oviducts, yolk glands, statocyst. 1 family.

ORDER LECITHOEPITHELIATA

Freshwater and marine; with complex pharynx opening at or near anterior end of body; ovovitellaria producing both egg and yolk cells. 2 families.

ORDER PROLECITHOPHORA

Mainly marine; with bulbous, tubular pharynx, simple sac-like gut, common genital opening (in some families with mouth also included to form a single orogenital pore), yolk cells surrounding ova, flagellate sperm. 7 families.

ORDER NEORHABDOCOELA

Marine, freshwater and terrestrial; with complex bulbous pharynx, straight non-ciliated gut, separate ovaries and yolk glands (or as separate portions of a single organ). Some commensal and parasitic spp. 32 families.

ORDER TEMNOCEPHALIDA

Commensal, freshwater; on gills, in branchial chamber or on surface of crustaceans, or on snails or turtles. Body flattened, with posterior adhesive disc and anterior finger-like 'tentacles', both disc and tentacles used in looping locomotion; without cilia over surface. Order sometimes raised to class rank; sometimes demoted to suborder of Neorhabdocoela. 2 families.

ORDER PROSERIATA

Interstitial, marine (and a few freshwater spp.); with tubular pharynx hanging in large pharyngeal cavity, unbranched gut, yolk glands, characteristically-formed statocyst; commonly with adhesive structures posteriorly. 8 families.

ORDER TRICLADIDA

Marine, freshwater and terrestrial; with large, often elongate, body, sometimes with adhesive organs. Tubular, folded pharynx in large cavity; intestine divided into single anterior and 2 posterior diverticulated branches; with long oviducts, many yolk glands, and, typically, with copulatory organs opening through common pore. 'Planarians'. 9 families.

ORDER POLYCLADIDA

Large, oval, often brightly coloured, greatly flattened, marine; with folded pharynx, intestine with numerous lateral branches, many eyes, many gonads, cephalic tentacles; without statocysts; sperm have 2 flagella with '9 + 1' microtubular ultrastructure. 29 families.

Class Monogenea

Flat, leaf-shaped flukes, mostly ectoparasitic on fish. Eggs hatch into onchomiracidia larvae which attach to host and then change gradually into adult form; no asexual multiplication phase whilst larval. Body with (often multiple) organs of attachment, including suckers, hooks and clamps, at both anterior and posterior ends, with covering of cuticle, and paired anterolateral excretory pores and well-developed gut.

ORDER MONOPISTHOCOTYLEA

Paired organs of attachment anteriorly and a single posterior attachment organ; without a genitointestinal canal. 14 families.

ORDER POLYOPISTHOCOTYLEA

Single oral sucker or a pair of eversible buccal suckers; 2 or more posterior attachment organs; with a genitointestinal canal. 28 families.

Class Trematoda

Endoparasitic flukes, mostly in vertebrates, with a complex life history involving 2 or more hosts (the first intermediate host characteristically a mollusc) and several larval stages, one or more of which multiplying asexually. Cylindrical or leaf-shaped body with well-developed gut, single posterior excretory pore, cuticle-covered surface, and 1 or 2 suckers.

Two subclasses distinguished: the Digenea (containing 5 out of the 6 trematode orders) and the Aspidogastrea (containing the single order, Aspidobothria).

ORDER ASPIDOBOTHRIA

Cylindrical, without attachment organs anteriorly; with either a very large, compartmented sucker covering whole ventral surface, or with ventral row of suckers. Life-cycle simple with 1 or no intermediate host. 3 families.

ORDER STRIGEIDIDA

Digeneans with thin-walled excretory vesicle; large miracidium larva with 2 pairs of flame cells; circaria larva with forked tail, penetration glands, and caudal excretory vessels. 29 families.

ORDER AZYGIIDA

Digeneans with thin-walled excretory vesicle; miracidium larva with 1 pair of flame cells, spinous and non-ciliated; large circaria larva eaten by second intermediate or definitive host. 26 families.

ORDER ECHINOSTOMIDA

Digeneans with thin-walled excretory vesicle; miracidium larva with 1 pair of flame cells; large-bodied circaria larva with single tail, numerous cystogenous glands, usually no penetration glands, and with caudal excretory vessels during development. 28 families.

ORDER PLAGIORCHIIDA

Digeneans with thick-walled, epithelial excretory vesicle; miracidium larva with 1 pair of flame cells; circaria larva with single tail, often reduced or absent, without caudal excretory vessels at any stage. 53 families.

ORDER OPISTHORCHIIDA

Digeneans with thick-walled, epithelial excretory vesicle; miracidium larva with 1 pair of flame cells; spinous circaria larva with single tail, often reduced in size, sometimes absent, and with caudal excretory vessels during development. 10 families.

Class Cestodaria

Flatworms without mouths or guts, endoparasitic in fish and turtles. Cuticle-covered body with single set of male and female organs, sometimes with anterior suckers, never with a scolex. First larval stage an onchosphere with 10 hooks.

ORDER AMPHILINIDA

Body with poorly-developed anterior holdfast and posterior genital pores. 2 families.

ORDER GYROCOTYLIDA

Body with simple anterior holdfast and posteriorly a cylindrical or funnel-shaped, frequently frilled, opening; genital pores anterior. 1 family.

Class Cestoda

Endoparasitic flatworms in guts of vertebrates as adults and in a variety of intermediate hosts as larvae. Body without mouth or gut; with anterior attachment organ (the scolex) bearing hooks, suckers, bothria, etc., a short neck region, and a strobila—a linear series of proglottids budded from the neck, or, if neck missing, from the scolex. Each proglottid develops set of reproductive organs and matures as it is distanced from neck by younger proglottids; eventually filling with eggs. Eggs develop into onchosphere larvae with 6 hooks. 'Tape worms'.

ORDER CARYOPHYLLIDA

Parasites of freshwater fish when adult and of oligochaete worms when larval; with poorly-developed scolex and undivided strobila with single set of reproductive organs (1 ovary, many testes). Probably neotenous larvae of unknown group of tape worms. 1 family.

ORDER SPATHEBOTHRIIDA

Parasites of fish when adult; scolex variably developed, sometimes absent, strobila not divided but with several serially-arranged sets of reproductive organs. Probably neotenous pseudophyllidan cestodes. 3 families.

ORDER TREPANORHYNCHIDA

Parasites of Chondrichthyes when adult; scolex with 2–4 muscular bothridia and, apically, with 4 hook-bearing tentacles, retractile into sheaths; strobila with numerous proglottids. 16 families.

ORDER PSEUDOPHYLLIDA

Parasites of all groups of vertebrates when adult; scolex with dorsal and ventral bothria, often fused for all or part of their length, sometimes with hooks; strobila with numerous proglottids, although divisions between proglottids poorly marked in some spp. 10 families.

ORDER LECANICEPHALIDA

Parasites of Chondrichthyes when adult; scolex divided into 2 regions by transverse groove, anterior part forming cushion-like organ, posterior region with 4 simple suckers, strobila with many proglottids. 4 families.

ORDER APORIDA

Parasites of ducks, geese and swans when adult; scolex with simple suckers or grooves and armed rostellum, strobila divided into proglottids only internally, no external signs of division; no genital ducts or pores, testes surrounding ovary. 1 family.

ORDER TETRAPHYLLIDA

Parasites of Chondrichthyes when adult; scolex with 4 pedunculate or sessile bothridia, often divided into loculi, sometimes with hooks, spines or suckers; strobila with many proglottids. 4 families.

ORDER DIPHYLLIDA

Parasites of Chondrichthyes when adult; scolex with peduncle, 2 spoon-shaped bothridia lined with small spines, either an apical organ or large rostellum with T-shaped hooks on apex; strobila cylindrical, with few proglottids. 2 families.

ORDER LITOBOTHRIDA

Parasites of Chondrichthyes when adult; scolex with single well-developed sucker; strobila with many proglottids, anterior proglottids cruciform in section, broad. 1 family.

ORDER NIPPOTAENIIDA

Parasites of freshwater teleosts when adult; scolex simple, rounded, with single apical sucker; strobila with few proglottids, round or oval in section. 1 family.

ORDER PROTEOCEPHALIDA

Parasites of fish, amphibians and reptiles when adult; scolex with 4 muscular suckers and sometimes with apical sucker or rostellum; strobila with distinct proglottids. 1 family.

ORDER CYCLOPHYLLIDA

Parasites of tetrapod vertebrates when adult; scolex with 4 suckers and, typically, apical rostellum armed with hooks or spines; strobila often with many proglottids, with 4 longitudinal excretory canals, with single compact vitellarium. 14 families.

Phylum Mesozoa

Mesozoans are a small group of enigmatic, ciliated, bilaterally symmetrical endoparasites composed of small numbers of cells and without any muscular, nervous, digestive, respiratory, excretory or skeletal systems. The vermiform solid body comprises two layers of cells: an outer layer of somatic cells enclosing one or more large axial germinative cells, which may produce and contain intracellular axoblast cells. Life histories are complex and involve an alternation of sexual and asexual generations. Some authors regard the Mesozoa as secondarily simplified Platyhelminthes; others consider them a multicellular group separately evolved from the protists.

The 50 known species are distributed between 2 classes (which may be unrelated).

Class Rhombozoa

Parasites in the kidney of cephalopod molluscs, with 20–30 somatic cells enclosing long, cylindrical axial cell containing 1 to > 100 axoblasts. Life-cycle only partly known: young cephalopods infected by infusoriform larva which migrates to kidney and develops into adult rhombozoan; whilst cephalopod still juvenile, axoblast cells produce vermiform larvae developing into nematogens which multiply asexually; when population density of nematogens in kidney becomes high, nematogens produce sexual rhombogens; rhombogens produce gametes which, on fertilization within the rhombogen axial cell, develop into infusoriform larvae which leave the rhombogen and the cephalopod.

ORDER DICYEMIDA

Nematogen with separate, ciliated, somatic cells; with 8 or 9 calotte cells comprising a head region. 1 family.

ORDER HETEROCYEMIDA

Nematogen with non-ciliated, syncytial, somatic cells; without a head region. 1 family.

Class Orthonectida

Tissue parasites of marine invertebrates in asexual phase; free-living in sexual generation. Free-living orthonectidans vermiform, annular, with somatic cells enclosing either male or female gametes; sperm discharged and penetrate female, ciliated larva forming after fertilization; larvae penetrate host organism, lose their ciliated somatic cells, become syncytial forming plasmodia, multiply asexually to spread through host tissues and produce sexual generation. 2 families.

Phylum Gnathostomula

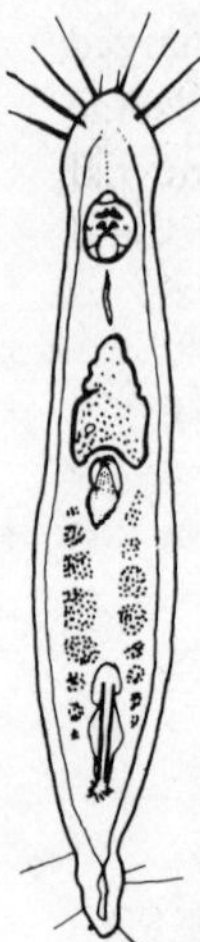

Gnathostomulans are small, bilaterally symmetrical, vermiform acoelomates of interstitial and detrital marine habitats. They are superficially similar to turbellarian Platyhelminthes, and, like them, have a blind-ending tubular gut, no circulatory or respiratory systems, no body cavity, a hermaphrodite reproductive system and a ciliated epidermis. Each epidermal cell, however, bears only a single cilium (a feature shared only with the Gastrotricha); the well-developed muscular pharynx bears paired jaws and a hard comb-like basal plate; the parenchyma is poorly-developed; and the anterior end, which may be demarkated from the trunk by a constriction, bears either a long rostrum or a series of sensory bristles, pits and cilia. The body, which may be very slender, is up to 1 mm long.

Some 100 species, within 2 orders, have been described to date; many more species probably await discovery.

ORDER FILOSPERMIDA

Sperm filiform; body without vagina, bursa or penis; without paired sensory organs anteriorly; with slender pointed rostrum, and paired apophyses on jaws. 2 families.

ORDER BURSOVAGINIDA

Sperm not filiform; with penis, bursa and, often, vagina; with paired sense organs anteriorly; without wing-shaped apophyses on jaws. 8 families.

Phylum Nemertea

Nemerteans, nemertines or rhynchocoels are vermiform, typically elongate animals similar in organization to the turbellarian Platyhelminthes, i.e., they lack a body cavity, have a ciliated epidermis, are unsegmented and more or less dorsoventrally flattened. In contrast, however, they have a gut with both mouth and anus, a blood vessel system in the parenchyma, and a characteristic eversible proboscis housed in a tubular cavity, the rhynchocoel, lying dorsally to the gut. The nervous system comprises cerebral ganglia and longitudinal cords. Most species are dioecious, oviparous, and possess excretory organs.

Although one group is terrestrial, a few species inhabit freshwater, and some are free-swimming in the sea; most nemertines are benthic and marine. Almost all are active carnivores using the proboscis to capture prey (the terrestrial species also use the proboscis in locomotion). Nemertines include the longest known animals (the unravelled portion of an only half-unravelled *Lineus longissimus* measured 30 m).

The 900 described species are placed in 2 classes and 4 orders.

Class Anopla

With separate mouth and proboscis pore; mouth below or behind cephalic ganglion; proboscis of uniform construction throughout its length and unarmed (except, in some spp., with rhabdite-like barbs); gut simple; nervous system within body wall.

ORDER PALAEONEMERTEA

Dermis gelatinous or absent; 2 or 3 layers of muscle in body wall (if 2, outer layer circular, inner longitudinal; if 3, outer and inner circular, middle longitudinal), with nervous system in inner longitudinal layer or beneath epidermis. Benthic marine. 4 families.

ORDER HETERONEMERTEA

Dermis well-developed; 3 layers of muscle in body wall (outer and inner longitudinal, middle circular), with nervous system between outer longitudinal and middle circular layers. Mostly benthic marine, some freshwater. 5 families.

Class Enopla

With mouth and proboscis opening through common aperture, or separately from a common atrium, in front of cerebral ganglion; nervous system internal to body wall musculature, which possesses outer circular and inner longitudinal layers.

ORDER HOPLONEMERTEA

Proboscis differentiated into regions, armed with one or more needle-like stylets; complex gut with lateral diverticula. Mostly benthic or pelagic marine, but also freshwater, terrestrial, commensal and parasitic spp. 28 families.

ORDER BDELLONEMERTEA

Leech-like, commensal within mantle cavity of marine bivalve molluscs, filter-feeder on host's water currents; proboscis simple, unarmed, opening into foregut; gut with ciliated papillae anteriorly, without lateral diverticula; body with posteroventral sucker. 1 family.

Phylum Gastrotricha

Gastrotrichs are small (< 4 mm long), free-living, unsegmented, bilaterally symmetrical, vermiform aquatic animals with an oval to elongate, dorsoventrally flattened body ending in a thin tail, a fork or in a rounded shape. They were once believed to possess a pseudocoelomic body cavity, but this is now known to be an artefact and gastrotrichs are acoelomate; also, like turbellarian Platyhelminthes, they lack respiratory and circulatory systems.

The body wall comprises a thin, chitinless cuticle, an epidermis, and longitudinal and circular muscle layers; the dorsal surface bears a monociliar covering, and cilia are also present ventrally in various patterns, being used in the creeping form of locomotion. Dorsally and laterally, the body may bear scales, spines or hooks, and 2–250 adhesive tubes which secrete a material for temporary attachment to the substratum. The gut possesses both mouth and anus, and a muscular pharynx; the nervous system comprises a large cerebral ganglion, longitudinal cords, and anterior sense organs (photoreceptors, bristles, cilia, and sensory pits); and freshwater species typically possess a pair of protonephridia. Most species are hermaphroditic, although some have adopted parthenogenesis. Cleavage of the egg is bilaterally radial; no distinct larval stage occurs.

The 450 known species are included in a single class with 2 orders.

ORDER MACRODASYIDA

Marine; strap-shaped body with adhesive tubes anteriorly, posteriorly and laterally; pharynx with pores; hermaphroditic. 6 families.

ORDER CHAETONOTIDA

Freshwater and marine; fusiform body with adhesive tubes only posteriorly, if at all; pharynx without pores; hermaphroditic or parthenogenetic. 7 families.

Phylum Rotifera

Rotifers are small (< 2 mm long), mostly free-living, unsegmented, bilaterally symmetrical aquatic animals, leading a swimming, crawling or sessile existence in freshwaters and, less commonly, in the sea. The body is elongated, trumpet-shaped or spherical and bears anteriorly a characteristic wheel-organ, or corona, a crown of cilia used in locomotion and/or feeding; posteriorly, the body ends in a foot (in all but planktonic spp.) which is used for attachment to the substratum. The body wall does not contain muscle layers, the muscles being organized in scattered bundles, and is covered by an intracellular chitinous cuticle which may be thickened into a case or lorica; a small pseudocoelomic body cavity surrounds the internal organs.

A gut is present with mouth, anus and a muscular jaw-bearing pharynx, the jaws being formed of seven elements; the nervous system comprises an anterior ganglion, two main longitudinal cords and sensory eyespot, antennae, bristles, etc.; one pair of protonephridia occurs; circulatory and respiratory systems are lacking. The organs of adult rotifers have a constant number of cells (or nuclei, since many organs are syncytial). The sexes are separate, although the males, which, except in one small group, are always smaller than the females, may be lacking and the rotifers are then parthenogenetic. Cleavage of the egg is spiral; there is no distinct larval stage.

The 1800 known species are apportioned between 3 classes.

Class Seisonidea

Large, marine; living on gills of the crustacean *Nebalia*; body elongate with vestigial corona; males fully developed. Only 1 genus.

Class Bdelloidea

Mainly freshwater; body vermiform, can be contracted by telescoping; corona in form of 2 wheels; 2 ovaries; retractable foot with up to 4 toes; parthenogenetic. A single order (Bdelloida), 4 families.

Class Monogonata

Body form variable; single ovary; male, if known, smaller than female, with single testis; foot with no more than 2 toes. Contains majority of rotifers.

ORDER PLOIMA

Mostly planktonic; corona in form of circumapical band; foot, if present, with 1–2 toes and pedal glands. 14 families.

ORDER FLOSCULARIIDA

Sessile and free-swimming; circumapical band of corona broken into trochal and cingular portions; foot absent or, if present, without toes, with numerous pedal glands. 4 families.

ORDER COLLOTHECIDA

Sessile, inhabiting gelatinous tubes; anterior end expanded into funnel with scalloped lobes bearing bristles or cilia; posterior end with long foot and terminal attachment disc. 1 family.

Phylum Kinorhyncha

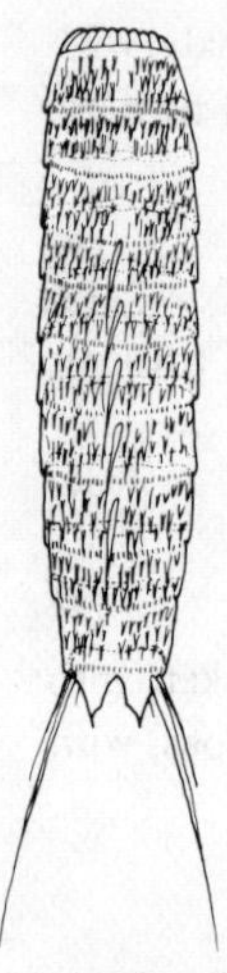

Kinorhynchs are small (< 2 mm long), squatly vermiform, free-living, bilaterally symmetrical, aquatic animals with a body divided externally into 13 segments, although with no internal segmentation, and with a well-developed pseudocoelomic body cavity derived from the persistent blastocoel. The body lacks cilia and is covered by a chitinous cuticle divided into plates with flexible articulating regions between.

The first segment forms the head, which bears a terminal mouth cone with anteriorly directed styles, and rows of recurved spines. The head, or first two segments, can be retracted into the body; the burrowing form of locomotion being effected by everting the head, the spines of which grip the sediment, and by pulling the trunk forwards by retracting the head again. The second ('neck') and/or third segments bear plates which can be closed over the second and/or first segments when they are retracted. The 11 trunk segments may bear long, curved, hollow, movable, backward-directed spines; one pair of adhesive tubes typically occurs anteroventrally and numerous such tubes may occur laterally.

The gut is tubular with mouth, anus and muscular pharynx; the nervous system comprises a multilobed brain, longitudinal cords running ventrally, bearing segmentally-arranged ganglia, and sensory bristles and, sometimes, pigment-cup photoreceptors; one pair of protonephridia is present; respiratory and circulatory systems are lacking. The sexes are separate and fertilization is internal.

The 100 known species inhabit marine sediments and are placed in 2 orders.

ORDER CYCLORHAGIDA

Head retractable into neck and trunk segments, plates on neck closing the opening; mid-dorsal, lateral and caudal spines present; trunk segments with cuticular hairs or denticles. 4 families.

ORDER HOMALORHAGIDA

Head and neck retractable into trunk segments, plates on 3rd segment closing the opening; without mid-dorsal or lateral spines, with caudal spines in only 1 genus; without cuticular hairs. 2 families.

Phylum Priapula

Priapulans are small to large, free-living, bilaterally symmetrical, unsegmented though annular, vermiform, benthic marine animals with short cylindrical bodies divided into two or three clearly-defined regions: a barrel-shaped proboscis used in locomotion and/or feeding and which can be wholely or partly withdrawn by invagination; an annulated trunk bearing warts, papillae or spines; and, in some species, one or two tail appendages of unknown but probably diverse function. A well-developed body cavity is present, serving as a hydrostatic skeleton—this is considered by some to be a coelom bounded by peritoneum, and by others to be a persistent blastocoel.

The proboscis bears the mouth apically and circles and/or longitudinal rows of scalids form papillae, spines, or, in one group, stiff tentacles; the anus is borne on the trunk. The body is covered by a chitinous cuticle, moulted periodically, beneath which are epidermal, circular muscle and longitudinal muscle layers. The excretory and reproductive systems are partly combined, the excretory elements being protonephridia; the nervous system includes an anterior nerve ring and ventral cord; no separate respiratory or circulatory systems occur although coelomic/pseudocoelomic cells contain haemerythrin. The sexes are separate and fertilization is typically external; cleavage is radial and leads eventually to a loricate larva.

The 10 known species are placed in 2 classes.

Class Seticoronaria

Small (<3 mm long), tube-dwelling; anterior scalids form 2 series of tentacles; proboscis only partly retractable; circumanal crown of hooks; no tail appendages. Tentacles form feeding trap. 1 family.

Class Priapulida

Small to large (5–150 mm long), vermiform; proboscis scalids form papillae and hooks, not tentacles; proboscis fully retractable; scattered hooks may be present on trunk but not organized into rings; 0–2 tail appendages. Active carnivores moving through sediment. 2 families.

Phylum Acanthocephala

Acanthocephalans are parasites of vertebrate intestines when adult and of invertebrates, usually arthropods, when juvenile. They are bilaterally symmetrical, cylindrical, unsegmented, although often superficially annular animals with a body comprising a proboscis armed with sclerotized hooks, which can be retracted into a muscular sac and which is used to anchor the worm to the host's gut wall, and a trunk which is often papillate or spiny. The body can be from 1 mm to 1 m long and is covered in two cuticular layers: an extracellular fibrous epicuticle and an inner cuticle derived from the syncytial epidermis. A large pseudocoelomic body cavity is present; no mouth or gut occurs; the muscular, nervous and excretory systems are all reduced (excretory organs, when present, are a pair of protonephridia), and a blood system is absent, although a lacunar system in the body wall may be circulatory; there is no respiratory system. The sexes are separate and fertilization is internal; there are several larval stages during the life-cycle (acanthor, acanthella and cystacanth).

The 1000 species are accommodated in 3 orders.

ORDER ARCHIACANTHOCEPHALA

Proboscis with hooks in concentric circles, proboscis sac single- or double-walled; brain near middle of proboscis; trunk without spines; main lacunar vessels mid-dorsal and mid-ventral, or dorsal only. Parasites of terrestrial animals. 4 families.

ORDER EOACANTHOCEPHALA

Proboscis with radially arranged hooks, proboscis sac single-walled; brain near middle or anterior end of proboscis; main lacunar canals not median. Parasites mostly of freshwater teleosts. 4 families.

ORDER PALAEACANTHOCEPHALA

Proboscis with hooks in alternating radial rows, proboscis sac double-walled; brain near middle or posterior end of proboscis; main lacunar canals lateral. Parasites mostly of freshwater teleosts. 13 families.

Phylum Nematoda

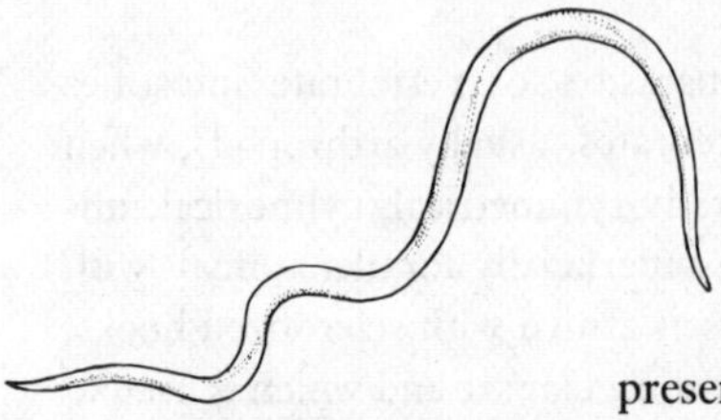

Nematodes or round-worms are unsegmented, bilaterally symmetrical, vermiform, free-living or parasitic animals with a cylindrical body usually tapering at both ends. They are present in almost all known habitat types, the free-living species being small (usually < 3 mm), whilst the parasitic ones may exceed 8 m. The body wall possesses complex cuticular layers (of collagen), a usually cellular epidermis, and a layer of longitudinal muscle, but no circular muscle; it encloses a large pseudocoelomic body cavity containing liquid under pressure and forming a hydrostatic skeleton. Without circular muscles and without cilia (except in sense organs), locomotion is via a series of S-shaped movements.

The tubular gut possesses a terminal mouth and subterminal anus, the mouth being surrounded by radially- or biradially-arranged lips and up to 16 setiform or papilliform sensory organs in 1–3 whorls; the anterior end also typically possesses two chemoreceptory amphids, often located near the mouth; the posterior end of the body, especially in parasitic species, may bear equivalent chemoreceptory phasmids. The excretory system, when present, is in the form of renette cells discharging anteroventrally; the nervous system includes an anterior nerve ring and a ventral double, ganglionated cord; respiratory and circulatory systems are absent. The sexes are separate, fertilization, where known, being internal; some species are parthenogenetic. Cleavage of the egg is determinate but not spiral; development is direct without larval stages.

The 15 000 known species are divided between 2 classes and 20 orders (more than 1 000 000 spp. probably exist).

Class Adenophorea

Mainly free-living; without phasmids; amphids located posteriorly in the head region; sensory bristles and papillae on the head and body; excretory organ, when present, single-celled and without collecting tubules; males without lateral extensions of the tail region.

ORDER ENOPLIDA

Predominantly marine; simple pouch-like amphids with slit-shaped or ellipsoidal apertures; well-developed cephalic sensillae in 3 whorls; body surface smooth or transversely striated. 13 families.

ORDER ISOLAIMIDA

Soil-dwelling; without amphids; mouth region with 6 hollow tubes and 2 whorls of papillae; body surface annulate anteriorly, with punctuations posteriorly. 1 family.

ORDER MONONCHIDA

Soil-dwelling and freshwater; small, cup-like amphids just posterior to lateral lips; cephalic sensillae in 2 whorls; buccal region with 1 or more massive teeth. Predators. 7 families.

ORDER DORYLAIMIDA

Soil-dwelling, freshwater, and parasites of plants; inverted stirrup-shaped amphids; 2 whorls of cephalic sensillae; buccal region with hollow spear or movable tooth; body surface smooth. 23 families.

ORDER TRICHOCEPHALIDA

Parasites of vertebrates; with spear in buccal region when juvenile, lost by adult stage; amphids near lips; stichosomous oesophageal glands; characteristically-arranged germinal zone in gonads. 3 families.

ORDER MERMITHIDA

Parasites of insects; long and slender; with pore-like or pocket-like amphids opening terminally; degenerate intestine forms storage organ. 2 families.

ORDER MUSPICEIDA

Parasites of vertebrates; without amphids, cephalic sensillae, or males; with reduced gut. 3 families.

ORDER ARAEOLAIMIDA

Mostly marine, but also soil-dwelling and freshwater; with spiral amphids; 3 whorls of cephalic sensillae; transverse striations on body surface. 11 families.

ORDER CHROMADORIDA

Mostly marine, but also soil-dwelling and freshwater; ornamented cuticles; crescentic valve in oesophagus; cephalic sensillae in 1 or 2 whorls; amphids variable (reniform, loops or multiple spirals). 9 families.

ORDER DESMOSCOLECIDA

Marine; conspicuously annular bodies frequently with bristles or scales; large circular amphids; reduced cephalic sensillae. 2 families.

ORDER DESMODORIDA

Mostly marine, but also freshwater; additional cuticular layers over head forming a helmet; annular body; sometimes with anterior and posterior adhesive tubes used in locomotion. 6 families.

ORDER MONHYSTERIDA

Marine, freshwater and soil-dwelling; 2nd and 3rd whorls of cephalic sensillae combined or only 3rd present; often with bristles on neck region and small teeth around mouth; amphids circular to spiral. 7 families.

Class Secernentea

Mostly terrestrial or parasitic; with phasmids; with amphids located anteriorly in the head region and opening on lateral lips; excretory system includes collecting tubules; without somatic setae or papillae, except sometimes on tail of males; males often with lateral extensions of tail region.

ORDER RHABDITIDA

Parasitic or soil-dwelling; 0–6 lips around mouth; tubular buccal region; tripartite oesophagus (swollen anterior; constriction; posterior bulb with valve); U-shaped excretory system. 22 families.

ORDER STRONGYLIDA

Parasites of vertebrates when adult; 3–6 lips around mouth; oesophagus cylindrical in adults; U-shaped excretory system; muscular bursa copulatrix in males. 14 families.

ORDER ASCARIDIDA

Parasites of vertebrates when adult; 3–6 lips around mouth; amphids pore-like; H-shaped excretory system. 23 families.

ORDER SPIRURIDA

Parasites of vertebrates when adult; 2 lips around mouth, sometimes 0 or 4; oesophagus with narrow anterior region and expanded (multinucleate) glandular posterior; males with long tails. 16 families.

ORDER CAMALLANIDA

Parasites of vertebrates when adult and of copepods when juvenile; without lips around mouth; uninucleate glands in posterior oesophageal region. 5 families.

ORDER DIPLOGASTERIDA

Soil-dwelling, a few parasitic in insects; short, setose cephalic sensillae; buccal cavity with movable teeth. 4 families.

ORDER TYLENCHIDA

Soil-dwelling and parasitic (in plants and animals); large buccal spear; anterior region of tripartite oesophagus divisible into procorpus and metacorpus, oesophageal glands opening into procorpus; asymmetrical excretory system with single collecting tube. 11 families.

ORDER APHELENCHIDA

Soil-dwelling and parasitic (in plants and animals); poorly-developed buccal spear; anterior region of tripartite oesophagus divisible into procorpus and metacorpus, oesophageal glands opening into metacorpus. 4 families.

Phylum Nematomorpha

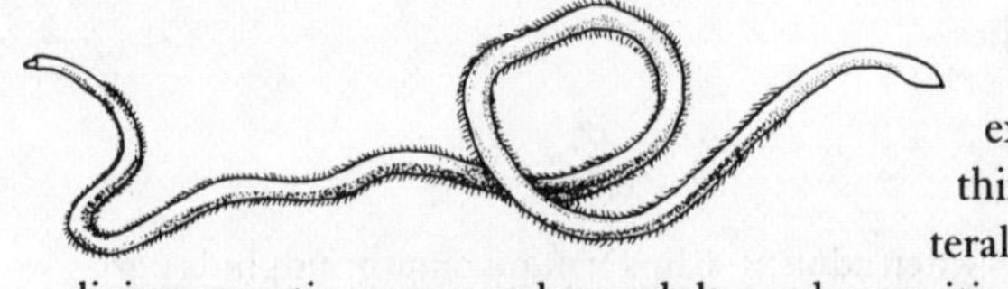

Nematomorphs are extremely long and thin, unsegmented, bilaterally symmetrical, free-living aquatic worms when adult, and parasitic in arthropods when juvenile. The basic organization of the body is similar to that of nematodes, i.e., the body wall comprises a collagen cuticle, an epidermis, and a layer of longitudinal muscle (circular muscles being absent), and it encloses a pseudocoelomic body cavity. In nematomorphs, however, the pseudocoelom is often occluded by mesenchyme, the alimentary canal of the adult is degenerate and non-functional—correlated with which the life of the free-living adult is short, serving only a reproductive function—the longitudinal nerve cord is single and non-ganglionated, and no excretory system is present. No respiratory or circulatory systems occur. The sexes are separate, and fertilization, where known, is internal. Cleavage of the egg is a modified spiral, and eventually produces a larval stage with stylets.

The 250 known species of 'horsehair worms' are divided between 2 orders.

ORDER NECTONEMATIDA

Marine; adults with double lateral row of setae along most of body length, used in swimming; open, unoccluded body cavity; males with single gonad; larvae and juveniles parasitize decapod crustaceans. 1 family.

ORDER GORDIOIDA

Freshwater or soil-dwelling; without natatory bristles; occluded pseudocoelom; paired gonads; thick cuticle; larvae parasitize insects. 5 families.

Phylum Sipuncula

Sipunculans are small (2 mm) to large (> 70 cm), bilaterally symmetrical, unsegmented, cylindrical, benthic, deposit-feeding, marine worms characterized by an elongate body divided into an anterior, long, narrow introvert completely retractable into a plumper posterior trunk, and by a large schizocoelic body cavity forming a hydrostatic skeleton. The introvert terminates in a mouth surrounded by ciliated tentacles or lobes which have their own hydraulic system of canals; coelomic pressure extends (unrolls) the introvert, and it is retracted by 1–4 muscles. Movement, even extension and retraction of the introvert, is slow.

The gut is basically U-shaped with a dorsal anus at the base of the introvert; the body wall is bounded externally by a cuticle and internally by a peritoneum, between which are an epidermis and layers of both circular and longitudinal muscle; no respiratory or blood systems are present although the coelomocytes contain haemerythrin; the excretory system is in the form of 1–2 metanephridia discharging through ventrolateral nephridiopores, and characteristic clumps of peritoneal cells termed urns; the nervous system includes a brain, circumoesophageal commissures, and a single, mid-ventral, non-ganglionated cord. With one hermaphrodite exception, the sexes are separate, a single gonad being present; gametes are shed into the coelom and discharged through the nephridiopores, and fertilization is external; the egg develops by spiral cleavage into a short-lived lecithotrophic trochophore and then, in several species, into a long-lived pelagosphaera; in some, development is direct.

The 350 living species are contained in a single class and order. 4 families.

Phylum Echiura

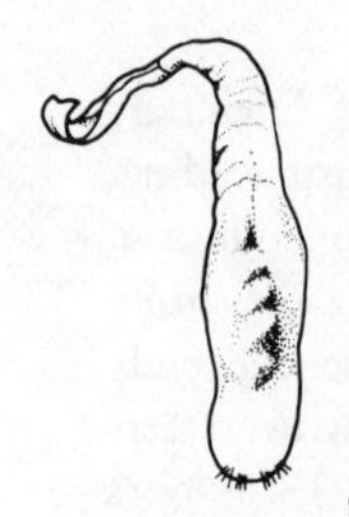

Echiurans are small to large, bilaterally symmetrical and unsegmented, benthic, deposit-feeding, marine worms with a cylindrical or bulbous body and a highly extensible (to > 2 m), but not eversible, anterior cephalic lobe (or 'proboscis'), down the ventral surface of which extends a ciliated groove. The proboscis, which is the food-collecting and gaseous-exchange organ, may be flat, spoon-shaped, long and slender with a forked tip, etc.; the plump body is characterized by a large schizocoelic body cavity, forming a hydrostatic skeleton, a pair of curved setae on the ventral surface just posterior to the mouth, and, often, a rough or papillate surface.

The long, coiled gut possesses both an anterior mouth at the base of the proboscis and a terminal, posterior anus; the excretory organs are a pair of anal sacs discharging into the posterior gut, with one to many metanephridia, usually in pairs; except in one order, a closed blood system occurs, with some cells containing haemoglobin; the nervous system includes a single, ventral, non-ganglionated cord and an anterior nerve ring; the body wall comprises cuticle, epidermis, circular, oblique and longitudinal muscle layers, and peritoneum. The sexes are separate, the diffuse gonads discharging gametes via the nephridiopores; fertilization is mainly external; cleavage of the egg is spiral, eventually producing a trochophore larva.

The 150 living species are placed in 3 orders, although the ordinal classification is little used in practice.

ORDER HETEROMYOTA

Longitudinal muscle layer external to circular and oblique layers; numerous (200–400), unpaired nephridia; proboscis very long; blood system closed. 1 family.

ORDER ECHIUROIDA

Longitudinal muscle layer between outer circular and inner oblique layers; up to 7 pairs of nephridia; blood system closed. 2 families.

ORDER XENOPNEUSTA

Longitudinal muscle layer between outer circular and inner oblique layers; 2–3 pairs of nephridia; proboscis reduced; blood system open; posterior gut enlarged forming respiratory organ, water pulsating through cloaca. 1 family.

Phylum Pogonophora

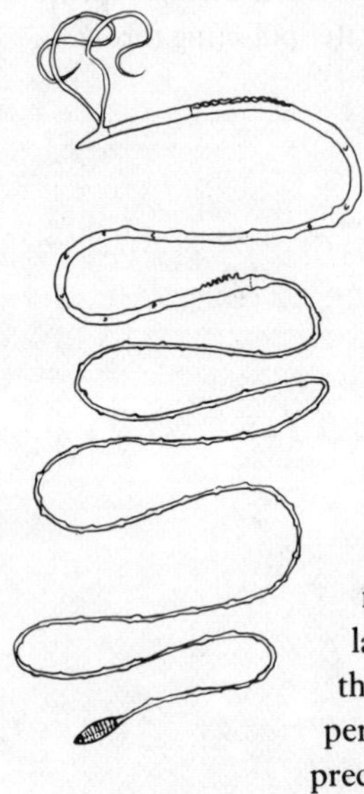

Pogonophorans are long (up to 2 m), slender, bilaterally symmetrical, tube-dwelling, benthic marine worms without mouth or gut and with a body divided into four regions—a cephalic lobe bearing beneath it from one to more than 1000 long, minutely pinnulate, ciliated tentacles; a short glandular region which secretes the tube; a long, often papillate trunk sometimes bearing setae; and a terminal, segmented opisthosome, usually with setae.

A closed blood system, containing haemoglobin, is present with extensions into the tentacles; the nervous system is within the epidermal basement lamina; each region of the body and each segment of the opisthosome possesses a body cavity which lacks a peritoneal lining but is not a persistent blastocoel, its precise nature is not yet known; the cavity of the cephalic lobe extends into the tentacles. The sexes are separate; the gonads are paired and cylindrical and located in the trunk; fertilization occurs outside the body but within the chitinous tube; the eggs cleave bilaterally and are brooded within the tube. Pogonophorans live in depths of greater than 100 m and are assumed to feed by absorbing dissolved organics, by micropinocytosis, and/or with the aid of symbiotic bacteria.

The 100 living species are divided between 2 classes.

Class Frenulata

The short glandular region has a raised ridge, the frenulum, running obliquely round it; trunk usually with setae; with 1 to > 200 tentacles; tubes anchored in soft sediments.

ORDER ATHECANEPHRIA

Cephalic lobe with sac-shaped body cavity connected to exterior by 2 lateral coelomoducts; sperm often in cylindrical spermatophores; with 1–20 tentacles. 2 families.

ORDER THECANEPHRIA

Cephalic lobe with horseshoe-shaped body cavity connected to exterior by 2 median coelomoducts; sperm usually in flat spermatophores; with 2 to > 200 tentacles. 4 families.

Class Afrenulata

Without frenulum on glandular region, replaced by 2 tissue folds meeting in dorsal mid-line and extending towards anterior end; plug of hardened tissue amongst tentacles used to close tube opening; without setae on trunk; with > 1000 tentacles; tubes attached to hard substrata.

A single order (Vestimentifera), 2 families.

Phylum Annelida

Annelids are short and microscopic to large and elongate, bilaterally symmetrical, mostly free-living, terrestrial and aquatic worms with a body comprising an anterior prostomium, a posterior pygidium and, between them, a series of metameric segments, budded from in front of the pygidium and each containing sets of serially repeated organs (at least, primitively). Typically, a large, open, schizocoelic body cavity occurs in each segment, functioning as a hydrostatic skeleton during movement; a tubular gut extends from the anteroventral mouth, located on the 1st segment (the peristomium), to the terminal anus on the pygidium; chitinous setae project from the body, the wall of which comprises layers of cuticle, epidermis, outer circular and inner longitudinal muscles, and peritoneum; and each segment is demarkated externally by annular furrows and internally by septa extending from body wall to gut.

The gut is usually a simple tube (absent in a few species), the pharynx being eversible in some, and sometimes with chitinous jaws and/or teeth; the excretory organs are proto- or metanephridia, typically with one pair per segment; the blood system is closed, with dorsal and ventral longitudinal vessels and segmental connections, some of which may function as hearts; the larger species have respiratory pigments (of a variety of types) dissolved in the plasma; the nervous system includes a prostomial brain, a pair of longitudinal, ventral cords, often fused together, with segmentally arranged ganglia connected by commissures, if not fused; complex sensory eyes and tentacles may be present. The sexes may be separate or hermaphroditic, with gonads in most or only a few segments; fertilization may be external or internal; the eggs cleave spirally to form a protostomatous embryo; larval stage, if it occurs, is a trochophore; asexual multiplication by fragmentation in some.

The 15 000 living annelid species are apportioned between 3 classes and 31 orders.

Class Polychaeta

Almost entirely marine; with lateral, essentially biramous, parapodia bearing setae in distinct fascicles, and, typically, with sensory and feeding organs on the pro- and peristomium which together form a head. Spacious coelom; often with reduced septa, thus cavities of adjacent segments confluent; pharynx may be everted a considerable distance. Sexes usually separate and fertilization external, gametes leaving body via simple ducts or rupture of body wall; trochophore larva commonly occurs. Pelagic or benthic; tube-dwelling or errant; free-living, commensal or parasitic.

ORDER ORBINIIDA

Elongate body with numerous segments, indistinctly divided into anterior dorsoventrally flattened region and posterior cylindrical part; finger-like dorsal gills; soft, sac-like proboscis; without prostomial appendages or palps. 2 families.

ORDER CTENODRILIDA

Very small, with few segments; without prostomial appendages, palps, gills or parapodial lobes; with few, simple setae arising directly from body wall. 2 families.

ORDER PSAMMODRILIDA

Small, interstitial; several segments in middle of body with elongate notopodia supported by acicula, parapodial lobes otherwise absent; at least 1 asetigerous anterior segment; without prostomial appendages or palps. 1 family.

ORDER COSSURIDA

Thread-like; numerous similar segments; single, median dorsal tentacle, often almost as long as body, arising on segment 4, 5, 6 or 7; parapodia biramous with very low lobes; without prostomial appendages. 1 family.

ORDER SPIONIDA

Small to medium sized; with 1 pair or 2 groups of longitudinally-grooved palps on peristomium or anterior segment; without prostomial appendages (except sometimes an occipital antenna) or jaws. 8 families.

ORDER QUESTIDA

Minute, slender, oligochaete-like; reproductive organs in only a few segments; simple setae arising directly from side of body; without prostomial appendages or cirri; non-eversible pharynx. 1 family.

ORDER CAPITELLIDA

Elongate, cylindrical; eversible thin-walled proboscis; parapodia poorly developed, without acicula or cirri, always biramous with some neuropodia forming transverse ridges; without prostomial appendages or palps. 3 families.

ORDER OPHELIIDA

Short-bodied, cylindrical, often annulate; eversible, thin-walled proboscis, short truncate neuropodia, finely tapering setae on all but 1st segment; without prostomial appendages or palps; sometimes with cylindrical lateral gills. 2 families.

ORDER PHYLLODOCIDA

Active predators or scavengers; muscular, cylindrical, eversible pharynx with 0–2 pairs of jaws; prostomium with 1(+) pair(s) of antennae; frontal or frontolateral palps sometimes present; well-developed parapodia with aciculae; simple and compound setae present. 28 families.

ORDER AMPHINOMIDA

Prostomium with 1–5 antennae and posterior caruncle; anterior segments encroach around and enclose prostomium and peristomium; cylindrical, rasping, pad-like proboscis; some parapodia with branched gills. 2 families.

ORDER SPINTHERIDA

Oval ectoparasites of sponges; cylindrical proboscis; prostomium with median antenna; parapodia with notopodial membraneous ridges, neuropodia with curved, composite hooks. 1 family.

ORDER EUNICIDA

Predatory, scavenging or parasitic; with 2–5 pairs jaws; eversible, extremely muscular proboscis; prostomium with or without appendages;

parapodia with well-developed neuropodia, reduced notopodia. 10 families.

ORDER STERNASPIDA

Short, broad, with few segments; posteroventral region with stiff, mineralized, chitinous shield; first 3 segments with rows of stout setae; posterior gills. 1 family.

ORDER OWENIIDA

Inhabit a sandy tube; with few, elongate segments, some indistinct and indicated only by their setae; prostomium fused to peristomium, often with lobes or frills; proboscis a muscular pad; dense fields of neuropodial hooks. 1 family.

ORDER FLABELLIGERIDA

Prostomium reduced and fused to peristomium, retractable into rest of body; body surface papillated, not divided into regions; parapodia biramous with reduced lobes; pharynx unarmed, non-eversible, with ventral pad; blood often containing chlorocruorin. 3 families.

ORDER POEOBIIDA

Pelagic, laterally compressed, elongate oval; 11 body segments, no external segmentation visible; body wall semitransparent, with thick gelatinous layer; without setae; head fused to body, retractable, with 1 pair of long tentacular palps; pharynx eversible. 1 family.

ORDER TEREBELLIDA

Tube-dwelling; peristomium with feeding tentacles; prostomium without appendages; body divided into 2 or 3 regions; with 1(+) pair(s) of gills. 5 families.

ORDER SABELLIDA

Tube-dwelling; reduced prostomium fused with peristomium, forming stiff, tentacular crown of gills; body divided into 2 regions, setal types in neuropodia and notopodia reversing at transition. 4 families.

ORDER NERILLIDA

Small, interstitial; few, transparent segments; ventral ciliated median groove; prostomium with 2 palps, 2 nuchal organs, 0–3 antennae; parapodia uniramous with cirrus and 1 fascicle of setae. 1 family.

ORDER DINOPHILIDA

Very small, interstitial; few, transparent segments; without appendages, parapodia or setae; uniformly ciliated on ventral surface, with bands or tufts of cilia on dorsal surface; sometimes with posterior adhesive glands. 1 family.

ORDER POLYGORDIIDA

Slender, elongate, mainly interstitial; many segments, 2 solid antennae, 2 nuchal slits, irridescent cuticle; without parapodia, often without setae or surface cilia. 1 family.

ORDER PROTODRILIDA

Slender, flattened, interstitial; transverse bands of cilia, with ventral ciliated groove; head with 2 long lateral tentacles, 2 nuchal organs; parapodia small or absent; setae often absent. 2 families.

ORDER MYZOSTOMIDA

Flat, disc-shaped, without distinct head, pygidium or external signs of segmentation; muscular pharynx; uniramous parapodia, each with 1 hooked seta, alternating with sucker-like organs; lateral cirri or papillae; parasitic in or on echinoderms, especially crinoids. Often regarded as a separate class. 7 families.

Class Aeolosomata

Small, interstitial or epizoic; without septa; with separate ventral nerve cords attached to epidermis; muscular prostomium ciliated ventrally and used as locomotory organ; few or no setae; often multiply asexually; hermaphroditic, with 1 ovary in middle of body, with several testes both anterior and posterior to ovary, with several pairs of unicellular spermathecae; spermatogenesis in body cavity, nephridia acting as male ducts. Aeolosomatans resemble interstitial polychaetes in some respects and oligochaetes in others: their affinities are still in doubt.

A single order (Aeolosomatida), 2 families.

Class Clitellata

Without parapodia or cephalic appendages; reduced or no setae; ganglia of nerve cords fused, cords not subepidermal; hermaphroditic, with reproductive organs confined to a few segments, with complex gonoducts; with glandular area of skin, the clitellum, which secretes a cocoon in which fertilized eggs enclosed; cross fertilization via copulatory organs; direct development. Common in soil and freshwaters; also in the sea. 3 subclasses recognized.

Subclass Oligochaeta

Without terminal sucker(s); segments not subdivided into annuli; coelom not occluded by botryoidal tissue; septa present; testes anterior to ovaries; setae typically present, though few; numbers of segments variable.

ORDER LUMBRICULIDA

At least 1 pair of male funnels in same segment as male pore(s) with which they are associated; freshwater. 1 family.

ORDER MONILIGASTRIDA

Testes and funnels in paired dorsal testis sacs suspended in a septum; each pair of male pores at posterior border of segment behind corresponding testis sac; terrestrial. 1 family.

ORDER HAPLOTAXIDA

Male funnels at least 1 segment anterior to that with pore(s) with which they are associated; terrestrial, freshwater and benthic marine. 23 families.

Subclass Branchiobdella

With sucker at each end of body; each segment divided into 2 annuli; coelom not occluded by botryoidal tissue except terminally; with septa;

testes anterior to ovaries; without setae; body of 15 segments, first 3–4 forming head and anterior sucker, last 1–2, the posterior sucker; < 1 cm long; commensal or parasitic on freshwater crayfish.

A single order (Branchiobdellida), 1 family.

Subclass Hirudinoida

With 1 or 2 suckers, posterior sucker always present; body of 30 or 34 segments; each segment divided into 2–14 annuli, usually 3–5; coelom occluded by botryoidal tissue; testes posterior to ovaries; setae always absent from all but anterior segments, usually completely lacking. Predators or parasites in freshwater, the sea and on land. 'Leeches'.

ORDER ACANTHOBDELLIDA

With paired setae on first few segments; body of 30 segments, each divided into 4 annuli; last 4 segments forming posterior sucker; without anterior sucker; coelom reduced though still evident and with septa; periodically parasitic on freshwater fish. 1 family.

ORDER RHYNCHOBDELLIDA

Without setae; body of 34 segments; last 7 segments forming posterior sucker; anterior sucker present; coelom reduced to lacunae; without blood pigment; without jaws; pharynx forming muscular protractile proboscis; freshwater and marine predators and parasites. 3 families.

ORDER ARHYNCHOBDELLIDA

Without setae; body of 34 segments; last 7 segments forming posterior sucker; anterior sucker present; coelom reduced and merged with blood system; haemoglobin present; jaws or stylets present; without proboscis; freshwater and terrestrial predators and parasites. 9 families.

Phylum Mollusca

Molluscs are small to very large, unsegmented, bilaterally symmetrical (sometimes distorted), essentially free-living animals with a body covered by a sheet of skin, the mantle, which in most species secretes an external, protective calcareous shell of 1–8 plates; secondarily, the shell may be incorporated within the body or lost, and, ancestrally, it may have been in the form of calcareous scales. No body cavity is present, although the open blood system includes large sinuses which may act as a hydrostatic haemocoel.

The basic molluscan body-plan has been modified markedly in a variety of directions by different phylogenetic lines. Characteristically, however, the body comprises: a head, often with sensory tentacles and eyes; a large, ventral, muscular, creeping foot; a dorsal hump containing the viscera; and posteriorly, between the mantle and the visceral mass, a cavity open to the exterior containing one pair of respiratory gills (ctenidia), the openings of the alimentary, nephridial and reproductive systems, and chemoreceptory sense organs. The gut is a relatively simple tube, the mouth with jaws and the buccal region with a tongue-like, toothed, chitinous ribbon, the radula, which can be protruded and has a characteristic rasping motion; the excretory system consists of one pair of metanephridial kidneys; the blood system includes a heart in a pericardial cavity; the nervous system includes several paired ganglia subject to variable degrees of fusion and a ladder-like ventral cord also subject to contraction. The sexes may be separate or hermaphroditic; one pair of gonads is typical; fertilization is basically external, although copulation and internal fertilization may occur; cleavage of the egg is spiral; a trochophore larval stage, and often a veliger, may occur, although these may be confined to the egg or lost. Molluscs vary from being dorsoventrally flattened to helically spiral, and from being vermiform to octopus-shaped; they are pelagic, benthic and, a few, even parasitic in the sea, and they are also successful on land and in freshwater.

Some 100 000 species are estimated to be extant, divided between 8 classes and 39 orders.

Class Monoplacophora

Small, deep-sea, almost bilaterally symmetrical; with single, cap- or cone-shaped dorsal shell; body with distinct head and radula, without eyes or sensory tentacles (except around mouth); foot weakly muscular; anus median, posterior; mantle cavity large, extending laterally and posteriorly around foot, with 5–6 pairs of ctenidia; 8 pairs of pedal–retractor muscles; 6 metanephridia; sexes separate; fertilization external.

A single order (Tryblidioida), 1 family.

Class Chaetodermomorpha

Vermiform, burrowing, cylindrical, marine; without a shell; mantle, with embedded imbricating scales, covers body; without foot; terminal mouth with cuticular plate; posterioterminal mantle cavity with 1 pair of ctenidia; without excretory organs, some without radula; nervous system with 3–6 pairs of ganglia, 2 pairs of longitudinal cords uniting posteriorly, dorsoterminal sense organ; sexes separate; inhabit vertical burrows, posterior end uppermost.

A single order (Caudofoveata), 3 families.

Class Neomeniomorpha

Vermiform, laterally compressed, marine; without a shell; with mantle enveloping body except for longitudinal, ventral pedal groove, mantle with 1(+) layer(s) of calcareous bodies; preoral sense organ in front of subterminal ventral mouth; ventrosubterminal posterior mantle cavity without ctenidia, sometimes with secondary gills; without excretory organs, some without radula; ganglia fused, with both ventral and dorsolateral longitudinal nerve cords; hermaphrodite; eat and often live on cnidarians.

ORDER APLOTEGMENTARIA

With 1 layer of calcareous bodies in mantle; thin cuticle without epidermal papillae. 7 families.

ORDER PACHYTEGMENTARIA

With several layers of needle-like calcareous spicules in mantle; thick cuticle usually with stalked epidermal papillae. 14 families.

Class Polyplacophora

Elongate or oval, dorsoventrally flattened, bilaterally symmetrical, marine; with dorsal shell of 8 transverse, serially overlapping plates embedded in, and sometimes covered by, a girdle; cuticle of girdle with spines, scales or bristles; large, muscular, ventral foot; poorly-differentiated head without eyes or tentacles; mantle cavity a groove around foot, with 6–88 pairs of ctenidia; anus subterminal; radula present; without jaws; sexes separate; mostly with larval stage. 'Chitons'.

ORDER LEPIDOPLEURIDA

Outer edge of shell plates without attachment teeth; mantle not extending over plates; few pairs of gills, gills posterior, near anus. 3 families.

ORDER ISCHNOCHITONIDA

Outer edge of shell plates with attachment teeth; mantle not extending over plates; gills occupying most of mantle groove, except near anus. 9 families.

ORDER ACANTHOCHITONIDA

Outer edge of shell plates with well-developed attachment teeth; mantle extending over plates; few gills along only part of mantle groove. 1 family.

Class Gastropoda

Asymmetrical; with single, often spirally coiled shell into which body can be withdrawn (shell lost or reduced in some); during development, visceral mass and mantle rotated relative to head and foot through 180°, so mantle cavity anterior and gut and nervous system in U-shape; head with tentacles and eyes, with jaws and radula; muscular creeping foot (modified in swimming and burrowing forms); mantle cavity with 1 pair of bipectinate ctenidia, often reduced or lost; hermaphroditic or with separate sexes; planktonic larvae or direct development; fertilization external or internal. 'Snails', 'slugs', etc. 3 subclasses are recognized.

Subclass Prosobranchia

Shell, torsion, and ctenidium/ctenidia retained; head with 1 pair of tentacles with eyes at their bases; usually with operculum; sexes usually separate; mainly marine.

ORDER ARCHAEOGASTROPODA

Shell with nacreous layer; radula with numerous teeth in transverse rows; 1–2 bipectinate ctenidia; mantle cavity without siphon; sexes separate, male without penis; nervous system not concentrated. 26 families.

ORDER MESOGASTROPODA

Shell without nacreous layer; radula with 7 teeth in each row; 1 (left) monopectinate ctenidium; mantle drawn out into siphon; sexes mostly separate, male with penis; nervous system concentrated. 96 families.

ORDER NEOGASTROPODA

Shell without nacreous layer; radula with 1–3 teeth in each row; 1 (left) monopectinate ctenidium; mantle forming siphon, carried within siphonal canal of the shell; sexes separate, male with penis; nervous system concentrated. 20 families.

Subclass Opisthobranchia

Marine; with various degrees of detorsion; shell reduced or absent; ctenidium and mantle cavity reduced or lost; hermaphrodite; usually without operculum; head with 1–4 pairs tentacles.

ORDER BULLOMORPHA

Usually with shell, rarely with operculum; radula and ctenidium usually present; head forming flattened shield for burrowing; foot with prominent parapodial lobes; penis unarmed. 15 families.

ORDER PYRAMIDELLOMORPHA

With well-developed, spirally-coiled shell; operculum present; long proboscis bearing pointed bristle for piercing prey; without radula; without gills. 1 family.

ORDER THECOSOMATA

With parapodial lobes of foot forming large, wing-like fins; cephalic tentacles well developed; shell and operculum present or absent; gill rarely present; planktonic. 'Shelled pteropods'. 5 families.

ORDER GYMNOSOMATA

With small, ventral parapodial fins; without shell; body with marked 'waist'; head with 2 pairs of tentacles; without mantle cavity; anterior gut with adhesive cones, arms or hooked structures protrusible through mouth; planktonic. 'Naked pteropods'. 7 families.

ORDER APLYSIOMORPHA

Reduced, internal or no shell; head elongate, with 2 pairs of tentacles and well-developed radula; large foot with parapodial lobes; small mantle cavity with gill. 'Sea hares'. 5 families.

ORDER PLEUROBRANCHOMORPHA

Internal, external or no shell; without mantle cavity but with large mantle overhanging body and large feathery gill on right side; head with 2 pairs of tentacles; radula present; large foot without parapodial lobes. 3 families.

ORDER ACOCHLIDIOIDA

Small, interstitial; without shell, mantle cavity, gill or parapodial lobes; calcareous spicules in skin; head with radula and 2 pairs of tentacles; visceral mass separated from, and often longer than, foot. 6 families.

ORDER SACOGLOSSA

Suctorial, with narrow radula with sharp piercing teeth, with buccal pump; external, internal or no shell, sometimes with 2 valves; body slug- or leaf-like, sometimes with dorsal cerata; head with 1 pair of tentacles. 7 families.

ORDER NUDIBRANCHIA

Without shell; body bilaterally symmetrical externally, slug-like, often with dorsal cerata and/or secondary gills; without ctenidia, mantle cavity or parapodial lobes; head with 2 pairs of tentacles, 1 pair with sheaths into which they can be withdrawn. 'Sea slugs'. 60 families.

ORDER ONCHIDIDA

Oval body without shell; extensive warty mantle covering head and body; large posterior respiratory lung opening near anus; head with 1 pair of tentacles with eyes on tips; radula present; penis near right tentacle. Sometimes regarded as belonging to the Pulmonata. 4 families.

Subclass Pulmonata

Mainly terrestrial and freshwater; mantle cavity forming lung with contractile opening; without ctenidia; with various degrees of detorsion; highly concentrated nervous system; hermaphrodite; without larval stages.

ORDER BASOMMATOPHORA

Aquatic; with shell; 1 pair of tentacles with eyes near their bases; tentacles retractile but cannot invaginate; sometimes with secondary gill. 17 families.

ORDER STYLOMMATOPHORA

Terrestrial; with or without shell; 2 pairs of tentacles with eyes on summits of hindmost pair; tentacles can invaginate. 'Slugs' and 'snails'. 56 families.

Class Bivalvia

Bilaterally symmetrical; laterally compressed body enclosed within 2 calcareous, lateral shells hinged dorsally by an elastic ligament and shell teeth, shells closed by 2 adductor muscles antagonistic to ligament; large mantle cavity, with posterior edges of mantle sometimes fused to form siphons; 1 pair of ctenidia, very large in most species and used for filter-feeding; greatly reduced head, without eyes or radula; mouth with palps; foot laterally compressed, often greatly reduced, in some forming burrowing organ; sexes usually separate; fertilization external; larval stages; aquatic, benthic, sedentary or sessile.

ORDER NUCULIDA

Shell valves similar, with numerous small hinge teeth, with equisized adductor muscles; ctenidia small, protobranch and purely respiratory; very large palp proboscides used for food-collection; foot large, with flat ventral surface. 7 families.

ORDER SOLEMYIDA

Shell valves thin, uncalcified along outer edges, without hinge teeth, anterior adductor muscle the larger; protobranch ctenidia, large, with leaf-like filaments on either side of central axis, used for respiration and feeding; palps small. 2 families.

ORDER ARCIDA

Shell valves heavy, similar, with numerous small hinge teeth, with equisized adductor muscles; filibranch ctenidia long, narrow, W-shaped in section, without interlamellar struts; without siphons; foot small. 7 families.

ORDER MYTILIDA

Shells wedge- or fan-shaped, with similar valves, reduced or absent hinge teeth, anteriorly terminal beak, reduced or absent anterior adductor muscle; filibranch ctenidia long, narrow, W-shaped in section, without interlamellar struts; without siphons; foot small, often secreting byssal threads. 2 families.

ORDER PTERIIDA

Shell valves dissimilar, often strongly sculptured, with wing-like processes, with few or no hinge teeth, reduced or absent anterior adductor muscle; 1 valve of shell often cemented to substratum; filibranch ctenidia long, narrow, W-shaped in section; without siphons; foot small. 15 families.

ORDER TRIGONIIDA

Shells trigonal, with similar valves, with 2–3 schizodont teeth in each valve, with small anterior adductor muscle; eulamellibranch ctenidia long, narrow, W-shaped in section, without interlamellar struts; without siphons. 1 family.

ORDER UNIONIDA

Shell with similar valves covered by well-developed periostracum, internally with nacreous layer, with schizodont hinge, with 2 adductor muscles; eulamellibranch ctenidia long, narrow, W-shaped in section, with interlamellar struts; ovoviviparous with parasitic larvae; freshwater. 'Freshwater mussels'. 7 families.

ORDER VENERIDA

Shell with similar valves, with 2 types of hinge teeth, with equisized adductor muscles; ventral mantle edges partially fused, usually with siphons; (usually) eulamellibranch ctenidia large, W-shaped in section, with interlamellar struts. 53 families.

ORDER MYIDA

Shell with thin, dissimilar valves, with few heterodont or no hinge teeth, with a reduced ligament, often with reduced anterior adductor muscle; ventral mantle edges fused (except where foot emerges), with long, large, often non-retractable siphons; eulamellibranch ctenidia W-shaped in section, with interlamellar struts. 8 families.

ORDER ANOMALODESMATA

Shell with dissimilar valves, with thick enrolled hinge, with equisized adductor muscles; fused mantle edges, with long, often fused, siphons sometimes enclosed in calcareous tube; ctenidia either eulamellibranch with interlamellar struts or modified to form muscular pumping septa (septibranch), mantle then site of gaseous exchange; hermaphrodite. Septibranch anomalodesmatans are sometimes separated as a distinct order. 12 families.

Class Scaphopoda

Bilaterally symmetrical, with an elongate body in tubular 1-piece shell, open at each end; tubular shell tapered, often curved, elephant-tusk shaped; mantle cavity large, extending along whole ventral surface, without gills; head without eyes, with radula, and paired clusters of clubbed contractile tentacles (capitula), head projecting from larger aperture of shell; foot cylindrical; sexes separate; fertilization external; benthic marine.

ORDER DENTALIDA

With numerous thin capitula; foot protactile, conical-tipped, with expandable lateral lobes. 4 families.

ORDER SIPHONODENTALIDA

With few, large, broad capitula; foot reduced to long, filiform process, with lateral lobes fused into frilled disc. 4 families.

Class Cephalopoda

Bilaterally symmetrical, with linearly-chambered shell, often reduced or lost; when external shell present, animal inhabits last chamber, a thin filament of living tissue (the siphuncle) extending through older chambers; head with large, complex eyes and circle of prehensile tentacles around mouth; mouth with radula and a beak; mantle muscular, large ventral cavity present with gills, with opening beneath head; reduced foot forming a funnel through which water forced by contraction of mantle, providing jet propulsion; sexes separate, some tentacles of male modified for copulation; benthic or pelagic marine. 2 subclasses.

Subclass Nautiloidea

Shell external, many chambered, coiled; head with many (80–90) suckerless tentacles, 4 modified in male for copulation; funnel of 2 separate folds; 2 pairs of ctenidia, 2 pairs of nephridia; eyes without cornea or lens; nervous system diffuse.

A single order (Nautiloida), 1 family.

Subclass Coleoidea

Shell reduced and internal or absent; head with 8 or 10 sucker-bearing tentacles, 1 pair being modified in male for copulation; funnel a single closed tube; 1 pair of ctenidia, 1 pair of nephridia; eyes with cornea and lens; nervous system large, concentrated.

ORDER SEPIOIDA

Shell internal, calcareous, often chambered; body short, broad, with lateral fins, with 8 short arms and 2 long arms bearing suckers only on spooned tips, retractile into pits. 'Cuttlefish'. 5 families.

ORDER TEUTHOIDA

Shell internal, reduced to horny 'pen'; body elongate, fusiform, with lateral fins, with 10 tentacles, 2 sometimes elongate but not retractile into pits. 'Squids'. 25 families.

ORDER OCTOPODA

Shell-less; 8 equisized tentacles joined by web of skin; body short, round, usually without fins; benthic. 'Octopi'. 12 families.

ORDER VAMPYROMORPHA

Shell reduced to thin, leaf-shaped, transparent vestige; 8 equisized tentacles joined by extensive web of skin, with 2 tendril-like, retractile filaments between 1st and 2nd pairs of tentacles; body short, plump, with large fins; pelagic, deep-sea. 1 family.

Phylum Phorona

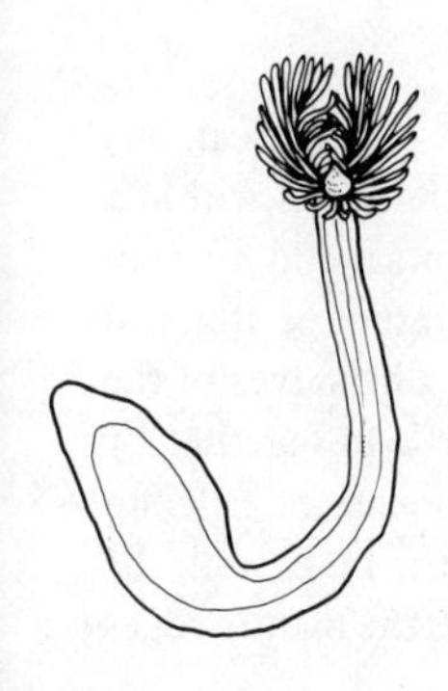

Phoronids are bilaterally symmetrical, vermiform, unsegmented, marine animals which secrete and inhabit a chitinous tube. Externally, the body bears no appendages or regional differentiation, except a terminal, horseshoe-shaped lophophore—a series of up to 500 hollow tentacles, each an extension of the body wall—around the mouth. Internally, the body is divided into three regions: a small preoral lobe or epistome overhanging the mouth; a short mesosome bearing the lophophore; and a long, cylindrical trunk containing most of the other organ systems. The mesosome and trunk (and possibly the epistome) possess a separate enterocoelic body cavity, that of the mesosome extending into the lophophore.

The body wall comprises an outer epithelium, a thin layer of circular muscle and a thick layer of longitudinal muscle, and an inner peritoneum; the gut is U-shaped with an anus near to, but outside, the lophophore; the excretory organs are a pair of metanephridia (protonephridia in the larva) in the trunk; the blood system is closed with haemoglobin in corpuscles; the nervous system is basiepithelial and mainly comprises a ring around the mouth and a single lateral nerve; the gonads are peritoneal aggregations of cells, and eggs and sperm are produced on opposite sides of the hermaphrodite body and escape via the nephridia; cleavage is radial and leads to an actinotroch larva; asexual multiplication also occurs; the adult is a sessile suspension-feeder (using the lophophore).

A single order (Phoronida), with only 2 genera and 10 species in 1 family.

Phylum Brachiopoda

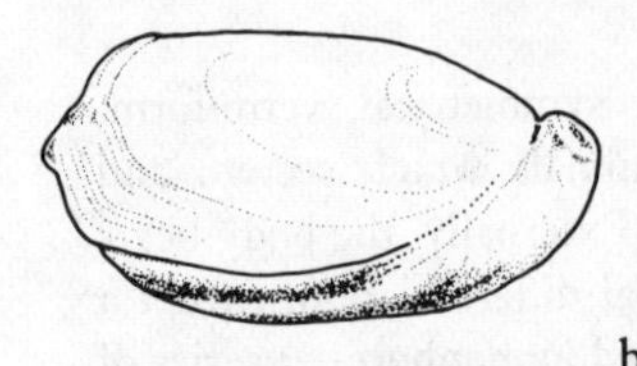

Brachiopods are bilaterally symmetrical, unsegmented marine animals enclosed within a bivalved shell, the valves dorsal and ventral, often cemented to the substratum or attached by a flexible stalk or pedicle. The valves of the shell are either hinged by a tooth and socket system or held together by muscles; the shell is composed either of calcium carbonate or calcium phosphate and chitin, and is secreted by extensions of the body wall, the mantle, canals of which often penetrate the shell punctae; the mantle edges often bear chitinous setae.

Most of the space within the shell valves is occupied by a large, circular, spiral or complexly-looped system of hollow tentacles, the lophophore, which is an extension of the anterior body wall; the lophophore is often supported by an internal calcium carbonate framework arising from the dorsal valve; the body itself is small and located posteriorly; the body and lophophore contain an enterocoelic or schizocoelic body cavity which extends into the mantle and pedicle; the gut is either blind-ending or possesses an anus opening into the mantle cavity; the blood system is open and without respiratory pigment, although some coelomocytes contain haemerythrin; the excretory system consists of 1–2 pairs of metanephridia; the nervous system includes a circumoesophageal ring from which nerves issue to lophophore, mantle, etc.; no discrete gonads occur, but two dorsal and two ventral aggregations of germ cells are associated with the peritoneum. Sexes are mostly separate and fertilization is usually external; the eggs cleave radially, form a protostomatous planktonic larva, and form mesoderm enterocoelically or by a modified form of enterocoely; asexual multiplication does not occur. 'Lamp shells'.

The 350 known species are placed in 2 rather distinct classes and 5 orders (over 12 000 fossil species are known).

Class Inarticulata

Shell valves, normally of calcium phosphate and chitin, held together solely by muscles; lophophore without supporting skeleton; gut with an anus; pedicle with muscles and coelomic cavity; body cavity forms schizocoelically from mesoderm produced enterocoelically.

ORDER LINGULIDA

Shell valves thin, identical, chitinophosphatic; with long, flexible pedicle emerging between the 2 shell valves; mantle setae forming siphon-like structures; occupy burrows in soft sediment. 1 family.

ORDER ACROTRETIDA

Small, circular in outline; often with ventral valve cemented to hard substrata; very long setae; pedicle absent or, if present, emerging through slot in ventral valve. 2 families.

Class Articulata

Shell valves calcium carbonate, with hinge of teeth and sockets; lophophore usually with supporting skeleton; gut without anus; pedicle without coelomic cavity or muscles; body cavity enterocoelic; pedicle, if present, emerging through slit or notch in ventral valve.

ORDER RHYNCHONELLIDA

Lophophore with 2 spirally-coiled lobes supported by spikes; shell not perforated by canals; pedicle present. 4 families.

ORDER TEREBRATULIDA

Lophophore with 2 simple lateral loops and coiled central lobe supported by a loop; shell perforated by canals; pedicle present or absent. 12 families.

ORDER THECIDEIDIDA

Lophophore supported by ridges on dorsal valve; ventral valve large, deep, cemented to substratum; dorsal valve small, lid-like, can be raised perpendicularly; pedicle absent. 2 families.

Phylum Bryozoa

Bryozoans are colonial aquatic animals, each colony comprising from a few to millions of zooids formed by the asexual division of an original founding ancestrula. Each zooid is an unsegmented, bilaterally symmetrical polyp with a circular or crescentic series of slender tentacles, the lophophore, a body or metasome, and, in some, a small preoral epistome; each zooid is in tissue contact with the other zooids surrounding it. Each body compartment possesses a body cavity, that of the metasome being spacious and connecting with that in the lophophore through a pore; the cavities are bounded by peritoneum but are neither enterocoels nor schizocoels, being formed *de novo* during metamorphosis of the larval stage. The epidermis of each zooid secretes a gelatinous, horny or calcareous shell enclosing the organism except for a region through which the lophophore can be extended (and retracted) by muscles, the aperture sometimes being protected by an operculum. Each zooid is small (some 0.5 mm), but the colonies may be large, varying in shape from flat encrustations and foliose, arboraceous structures to widely separate zooids connected by stolons; almost all are sessile. Within a colony, zooids may be polymorphic with individuals specialized for feeding, defence, brooding of the young, etc.

Each polyp typically possesses a U-shaped gut with a mouth within and an anus outside the lophophore; lacks excretory organs and a blood system; possesses a simple nervous system with a ring around the anterior gut from which individual nerves issue; and has peritoneal gonads, the gametes leaving via coelomoducts. Colonies are hermaphroditic, individual polyps being hermaphrodite or single-sexed; the eggs cleave radially, develop protostomatously and are often brooded, although a planktonic larval stage may occur.

Some 4000 species belong to the Bryozoa (also known as the Ectoprocta and Polyzoa), divided between 3 classes and 4 orders.

Class Phylactolaemata

Monomorphic, freshwater; large cylindrical zooids in non-calcareous colonies; polyps with body wall musculature, large horseshoe-shaped lophophore, well-developed epistome with a coelomic compartment, and resting buds (statoblasts) which overwinter, disperse and 'germinate' when conditions improve. Coelomic compartments of different individuals intercommunicate. Some colonies mobile.

A single order (Plumatellida), 4 families.

Class Stenolaemata

Marine; cylindrical zooids in tubular calcareous shells, circular terminal aperture without an operculum; without an epistome; circular lophophore not protruded by deformation of the calcareous covering but by hydrostatic pressure and muscular action; limited polymorphism; polyembryony is characteristic (primary embryo divides to form secondary embryos which themselves divide, etc. until more than 100 embryos formed).

A single order (Stenostomata) (often termed Cyclostomata, a name also used for an order of fish, see p. 249), 20 families.

Class Gymnolaemata

Polymorphic, mostly marine; with non- or partially-calcified walls, which when deformed protrude the lophophore; without epistome or body wall musculature; lophophore circular; lophophoral orifice closable; interconnecting pores in polyp walls plugged by special cells.

ORDER CTENOSTOMATA

Zooid walls membranous or gelatinous; lophophoral orifice closed by flap or collar; without ovicells, sometimes with brood chambers containing several embryos. 21 families.

ORDER CHEILOSTOMATA

Zooid walls partially calcified, deformation of frontal wall protrudes lophophore; lophophoral orifice closed by operculum; with ovicells; highly polymorphic. 86 families.

Phylum Entoprocta

Entoprocts are small, bilaterally symmetrical, unsegmented, solitary or colonial, mostly sessile aquatic animals with a rounded body (calyx) bearing aborally a cylindrical attachment stalk and orally a horseshoe-shaped ring of 8–36 tentacles which can contract and fold inwards. In contrast to the lophophorate phyla, water is drawn from outside the tentacles, between them, and then leaves from the centre of the tentacular ring; correspondingly, the anus of the U-shaped gut is located within the tentacular ring, together with the mouth.

The calyx contains one pair of protonephridia opening through a common pore, a small nerve ganglion from which individual nerves issue, an extensive body cavity—generally regarded as a pseudocoelom—extending into the tentacles and filled with gelatinous mesenchyme, and 1–2 pairs of gonads discharging through a common pore; the body wall lacks muscle layers and comprises an outer cuticle and an underlying epithelium; no blood system is present. Asexual multiplication by budding is common, in colonial forms new individuals are budded from a creeping stolon; most are hermaphrodite, internal fertilization, but not copulation, occurs; the egg cleaves spirally to form a trochophore-like planktonic larva. Most species are marine, living attached to hard substrata and other organisms; 1 genus is freshwater. Entoprocts have variously been allied with the other pseudocoelomate phyla and with the bryozoans but their affinities remain obscure.

The 150 living species are placed in 4 families; no orders are generally distinguished.

Phylum Crustacea

Crustaceans are small to large, bilaterally symmetrical, metamerically segmented animals with biramous limbs and a jointed exoskeleton which is moulted periodically to permit growth; they were once united with the uniramians and chelicerates in the phylum Arthropoda, but all three groups are now believed to have had independent origins. The body comprises a presegmental acron, a postsegmental telson, and an intervening series of segments; these are organized to form a head, thorax and abdomen. The head bears five pairs of appendages: antennules, antennae, jaws (mandibles) flanking the ventral mouth, and two pairs of accessory jaws (1st and 2nd maxillae); compound eyes are usually present. Each thoracic and abdominal segment bears one pair of appendages (those of the abdomen variably reduced or lost in some groups): each appendage comprises a basal protopodite and, arising from it, an inner endopodite and an outer exopodite; each of the three basic limb sections may bear projections, termed respectively epipodites, endites and exites, some of which may function as gills; the endo- and exopodites are multiarticulate. The segments of the head are fused together and, in many species, a fold of exoskeleton (the carapace) extends dorsally and laterally back from the head to cover some or all anterior segments.

The exoskeleton is chitinous and usually calcified; the gut is straight, possesses an anus at the base of the telson and a pair of diverticula forming a hepatopancreas; the blood system is open, with the blood sinuses forming the only body cavity, a dorsal heart and respiratory pigments in the plasma (haemocyanin or erythrocruorin) may be present; the excretory organs are glands comprising a closed end sac and a distal tubule located in one or more pairs of appendages (antennae, 2nd maxillae, etc.); the nervous system, which may show various degrees of concentration, includes a circumoesophageal ganglion system and a double ventral cord with a pair of interconnected ganglia in each segment. The sexes are usually separate, the gonads being paired, and copulation typically occurs; the eggs cleave in a highly modified spiral fashion, are often brooded, and usually hatch into a nauplius larva with three pairs of appendages (antennules, antennae and mandibles) and a median compound eye.

The 39 000 living species are divided between 9 classes and 38 orders.

Class Cephalocarida

Small (< 4 mm long); horseshoe-shaped head, 8-segmented thorax, 12-segmented abdomen; antennule uniramous; first 7 thoracic limbs triramous, with large flattened pseudepipodite; abdomen without appendages except caudal furca with long bristles; without carapace or eyes; hermaphrodite, with separate paired testes and ovaries discharging through common ducts; benthic marine. 2 families.

Class Remipedia

Small to medium sized, slender, elongate; dorsoventrally flattened body with short cephalothorax bearing cephalic shield and 7 pairs of appendages, and with long abdomen of 32 undifferentiated segments, each, except last, bearing laterally-directed, biramous, paddlelike, setose swimming appendages; all abdominal appendages similar; with caudal furca; without eyes; with 1 pair of pre-antennular frontal rod-like processes; antennules and antennae small, biramous; maxillules, maxillae and maxillipeds uniramous, prehensile; mandibles palpless; maxillipedal segment fused to cephalon; recently discovered in marine cave.

The class contains a single species.

Class Branchiopoda

Generally small; with short body, reduced unsegmented antennule, leaflike thoracic limbs with epipodal gills; maxillae small, 2nd maxillae often absent; without abdominal appendages except caudal furca; characteristic nauplius larva with numerous segments and further appendages added at each moult; carapace, when present, forms bivalved shell; compound eyes usually present; females often parthenogenetic, sexually produced, resistant eggs laid only when conditions adverse; mostly freshwater.

ORDER ANOSTRACA

Carapace absent; elongate body; stalked eyes; thorax usually with 11 limb-bearing segments; abdomen of 8 segments plus telson; telson with unsegmented furca. 'Fairy' or 'Brine shrimps'. 7 families.

ORDER NOTOSTRACA

Carapace forms a large dorsal shield; antennae reduced or absent in adult; each abdominal segment divided into several annuli. 'Tadpole shrimps'. 1 family.

ORDER CONCHOSTRACA

Carapace forms a bivalved shell which encloses body and large head; antennae large, used in locomotion; 10–28 pairs of thoracic appendages, decreasing in size towards posterior; caudal furca clawlike. 5 families.

ORDER CLADOCERA

Carapace valve single, but folded along longitudinal dorsal midline to form pseudobivalve shell enclosing body but not head, sometimes reduced to a brood-pouch; median, sessile compound eye; 4–6 pairs of thoracic appendages; posterior abdomen flexed forwards, with clawlike furca. 'Water fleas'. 11 families.

Class Ostracoda

Small; enclosed within bivalved carapace; body unsegmented externally, with 5–7 pairs of appendages; carapace shed and reconstituted at each moult.

ORDER MYODOCOPIDA

Paired, lateral compound eyes; notches in anterior carapace margin through which antennae can project; 7 pairs of appendages, including 2 pairs of thoracic limbs; biramous antennae with large basal segment; marine. 7 families.

ORDER CLADOCOPIDA

Small (< 1 mm); without eyes, heart, thoracic limbs or notch for antennae; 5 pairs of appendages; biramous antennae; benthic marine. 1 family.

ORDER PODOCOPIDA

Ventral margin of carapace valves concave; valves unequal; antenna uniramous, exopodite reduced to bristle; 2 pairs of thoracic limbs; compound eyes absent; marine, freshwater and terrestrial. 45 families.

ORDER PLATYCOPIDA

Small (<1.5 mm); asymmetrical with right valve overlapping left; antennae biramous; 1 pair of thoracic limbs, leaf-like in males, pincer-like in females; benthic marine. 1 family.

ORDER PALAEOCOPIDA

Known only from empty (presumed Recent) valves, order otherwise extinct; valves elongate, subelliptical, with long straight hinge; dorsal margin straight, nearly as long as total length; contact margins with wide, septate frill; with 6 muscle scars in ring; marine. 1 family.

Class Mystacocarida

Minute (<1 mm), elongate; head divided into 2 regions; without compound eyes, with 2 pairs ocelli; 4 pairs of simple, singly-segmented appendages on thoracic segments 2–5; antennules, antennae and maxillae large; appendages of 1st thoracic segment form maxillipeds; abdomen 5-segmented plus telson with large, clawlike furca; pairs of dentate furrows laterally on posterior head, thorax and abdomen; benthic marine.

The class contains a single order (Derocheilocarida) with 1 family.

Class Copepoda

Usually small; short body composed of a head fused to 1+ thoracic segments and 8 free segments plus telson (with further fusion in some); without carapace or paired eyes, with well-developed nauplius eye; uniramous antennules, appendages of 1st thoracic segment forming maxillipeds, and 4–5 other pairs of thoracic appendages; body often with rounded anterior margin, oval head + thorax, narrow abdomen ending in caudal furca; commensal, free-living and parasitic, parasitic species highly modified.

ORDER CALANOIDA

Free-living; long antennules, biramous antennae; with articulation between thorax and abdomen; planktonic, mostly marine. 40 families.

ORDER CYCLOPOIDA

Free-living and parasitic; short antennules, uniramous antennae; with articulation between 4th and 5th free thoracic segments, lost in some parasitic forms; female carries 2 egg sacs; planktonic and benthic, marine and freshwater. Notodelphyoid families are sometimes separated to form a distinct order. 14 families.

ORDER MISOPHRIOIDA

Free-living; short prehensile antennules, biramous antennae; with articulation between 4th and 5th free thoracic segments; with a heart; female carries 1 egg sac; benthic marine. 1 family.

ORDER MORMONILLOIDA

Free-living, planktonic; elongate antennules of 3 or 4 long articles, biramous antennae; with articulation between 4th and 5th free thoracic segments; without heart, without 5th or 6th pairs of legs; marine. 1 family.

ORDER HARPACTICOIDA

Mostly free-living; small body tapering posteriorly, without marked constriction; very short antennules, biramous antennae; with articulation between 4th and 5th free thoracic segments; benthic and planktonic, marine and freshwater. 34 families.

ORDER MONSTRILLOIDA

Larvae parasitic in polychaetes and gastropods, adults free-living but non-feeding; adult without antennae, mouthparts or gut, usually with well-developed antennules; infective as nauplii. 2 families.

ORDER SIPHONOSTOMATOIDA

Ectoparasitic; with reduced antennules, prehensile uniramous antennae, piercing stylet-like mandibles in tubular siphon-like mouth; with articulation between 3rd and 4th or 4th and 5th free thoracic segments, if not lost; marine and freshwater. 42 families.

ORDER POECILOSTOMATOIDA

Ectoparasitic; antennules often well-developed, antennae uniramous; falcate mandibles in slit-like mouth; with articulation between 4th and 5th free thoracic segments, if not lost; marine and freshwater. 42 families.

Class Branchiura

Flattened ectoparasites of fish; carapace covers head and most of body; with 3 free thoracic segments, paired movable compound eyes, median ocellus, bilobed unsegmented abdomen; 1st maxillae form suckers; antennules with large terminal hook. Attachment to host periodic, biramous thoracic limbs used for locomotion. 1 family.

Class Cirripedia

Free-living or parasitic; adults sessile; parasitic forms highly modified. Free-living forms attached to substratum by cement from antennules; carapace covering body and limbs, often with discrete calcareous plates; reduced abdomen and segmentation; 6 pairs of biramous thoracic limbs (cirri) used to comb seston from water; adults without antennae or compound eyes; mostly hermaphrodite with cross-fertilization using elongate penis; marine. 'Barnacles'.

ORDER THORACICA

Free-living; adults permanently attached to substratum by stalk or disc-like base; calcareous plates; 6 pairs of thoracic limbs. 17 families.

ORDER ACROTHORACICA

Free-living, burrow into rock or calcareous shells; attach by disc; calcareous plates absent; sexes separate; males minute, attached to females; females with 4–6 pairs of thoracic limbs, 1st pair separated by gap from remaining pairs; without abdomen. 3 families.

ORDER ASCOTHORACICA

Ecto- and endoparasites of cnidarians and echinoderms; prehensile antennules, mouthparts forming a suction cone, thoracic limbs not concerned with feeding; abdomen present. 4 families.

ORDER RHIZOCEPHALA

Parasites of crustaceans (mainly decapods) recognizable as cirripedes only when larval; without appendages or segmentation; body sac-like, external, reproductive, connected by short stalk to internal network of food-absorbing 'roots'. 7 families.

Class Malacostraca

Small to very large; often with heavily calcified exoskeleton; 8 thoracic and 6 (rarely 7) abdominal segments; usually with stalked compound eyes and carapace; carapace never covering more than first 1 or 2 abdominal segments; first 1–3 pairs of thoracic appendages forming maxillipeds, 5 or more posterior pairs forming walking or swimming legs (pereiopods), 1+ pairs often being chelate; abdomen with appendages (pleopods) modified for swimming and reproduction (sometimes for respiration), last pair (uropods) broad, forming tail fan with telson; fertilized eggs often brooded.

ORDER LEPTOSTRACA

Large bivalved carapace with anterior hinged plate over head; abdomen 7-segmented, 7th segment without appendages, with long caudal furca; protopodite 3-segmented; marine. 1 family.

ORDER STOMATOPODA

Short carapace leaving 4+ thoracic segments free, with anterior movable plate covering eye stalks; triramous antennules; first 5 pairs of thoracic legs uniramous, 2nd pair large, subchelate; abdominal limbs with gills; marine. 'Mantis shrimps'. 12 families.

ORDER ANASPIDACEA

Carapace absent; 1st thoracic segment fused to head but marked by groove; all thoracic appendages similar; eggs not brooded; freshwater. 4 families (one of which is sometimes regarded as constituting the separate order Stygocaridacea).

ORDER BATHYNELLACEA

Carapace absent; 1st thoracic segment free; all thoracic appendages similar; eggs not brooded; at most, 2 pairs of pleopods; uropods not forming tail fan; without eyes; elongate, worm-like, interstitial; freshwater. 3 families.

ORDER THERMOSBAENACEA

Body cylindrical without constriction between thorax and abdomen; carapace covering first 4 thoracic segments; 1st thoracic segment fused to head; telson fused to 6th abdominal segment; without eyes; only 1 pair of maxillipeds; freshwater. 2 families.

ORDER MYSIDACEA

Carapace covering most of thorax, fused to first 3 thoracic segments; eye stalks movable, covered by rostrum; 1 or 2 pairs of maxillipeds; 6–7 pairs of pereiopods with filamentous exopodites; both antennal and maxillary excretory glands present; mainly marine. 6 families.

ORDER CUMACEA

Head and thorax large, abdomen narrow; carapace fused dorsally to first 3 or 4 thoracic segments, forming large lateral branchial chambers, extending forwards beyond head; eyes sessile, often fused together; 3 pairs of maxillipeds; 4 oostegites; marine. 8 families.

ORDER SPELAEOGRIPHACEA

Elongate, cylindrical; carapace fused to 1st thoracic segment, overlapping 2nd; long abdomen (>one-half body length); antennal flagellum almost length of body; 1 pair of maxillipeds; without eyes; 5 oostegites; freshwater, cavernicolous. 1 family.

ORDER TANAIDACEA

Small; often somewhat flattened; small carapace fused to first 2 thoracic segments, forming lateral branchial chambers; eye stalks immovable; 1 pair of maxillipeds; 2nd pair of thoracic limbs large, chelate; 3rd pair of thoracic limbs forms burrowing organs; abdomen short, uropods not forming tail fan; marine. 19 families.

ORDER ISOPODA

Body dorsoventrally flattened, without carapace; first 1 or 2 thoracic segments fused to head; with short abdomen, sessile eyes, 1 pair of maxillipeds, small uniramous antennules; pereiopods without exopodites, forming walking legs; pleopods often forming gills; free-living and parasitic; marine, freshwater and terrestrial. 100 families.

ORDER AMPHIPODA

Body typically laterally compressed, without carapace; 1st (rarely also 2nd) thoracic segment fused to head; with sessile eyes, 1 pair of maxillipeds; pereiopods without exopodites; gills and heart thoracic; first 3 pairs of abdominal appendages form pleopods, last 3 pairs uropod-like; free-living, a few ectoparasitic; marine, freshwater, a few terrestrial. 100 families.

ORDER EUPHAUSIACEA

Carapace fused dorsally to all thoracic segments, not covering gills; eyes stalked; without maxillipeds, all thoracic limbs similar, with exopodites and podobranchial gills; pelagic marine. 'Krill'. 2 families.

ORDER DECAPODA

Carapace fused dorsally to all thoracic segments, extending laterally to enclose branchial chambers; eyes stalked; with podo-, pleuro- and arthrobranchial gills; 1(+) pereiopod(s) often chelate; 3 pairs of maxillipeds; sometimes very large; marine, freshwater, terrestrial. 'Prawns', 'lobsters', 'crabs'. 106 families (one of which is often regarded as constituting the separate order Amphionidacea).

Phylum Chelicerata

Chelicerates are small to medium sized, bilaterally symmetrical and metamerically segmented animals with a body divided into two regions: a prosoma comprising the presegmental acron and the six anterior trunk segments; and an opisthosoma of the postsegmental telson and up to 12 abdominal segments. They were once united with the crustaceans and the uniramians in the phylum Arthropoda, but all three groups are now believed to have had independent origins. The chelicerate prosoma usually bears simple ocelli, rarely compound eyes, and six pairs of jointed, uniramous appendages: preoral chelicerae, often forming chelae; pedipalps which may be leg-like, antenna-like or chelate; and four pairs of walking legs. The opisthosoma bears appendages forming organs of gaseous exchange or spinnerets; it is often divided into an anterior mesosoma and a posterior metasoma.

The anterior four or all six prosomal segments may fuse with the acron; a chitinous exoskeleton is present and is moulted periodically to permit growth; the gut is straight and ends at an anus in front of the telson, diverticula are present; the blood system is open and blood sinuses form the only body cavity, a dorsal heart is present and haemocyanin sometimes occurs in the plasma; the respiratory organs include book gills (paired sets of flap-like lamellae) and the internal book lungs and tracheae derived from them; excretion is via blind-ending tubule systems discharging into the gut (Malpighian tubules) or at the bases of appendages (coxal glands); the nervous system is often highly concentrated from a basic circum-oesophageal ganglion plus ventral, double, segmentally-ganglionated cord into a circumoesophageal brain from which nerves issue directly. The sexes are separate, with single or paired gonads and external or internal fertilization, often via spermatophores; the eggs are centrolecithal and development does not include a larval form markedly distinct from the adult.

The 63 000 known species are divided between 3 classes and 15 orders.

Class Merostomata

Marine; heavy exoskeleton; prosoma with large horseshoe-shaped carapace, separated from opisthosoma by hinge; telson forms long tail spine; 2

lateral compound eyes and 2 median ocelli; chelicerae small; pedipalps leg-like, chelate; walking legs chelate, except last pair; last pair of walking legs with leaf-like processes used for burrowing; spiny gnathobases of limbs macerate food; 1st pair of opisthosomal appendages form flap over reproductive openings; 2nd–6th pairs of opisthosomal appendages form swimming and gaseous exchange organs; opisthosoma unsegmented, with lateral spines; excretory organs coxal glands; fertilization external; with a 'trilobite larva'; benthic. 'Horseshoe crabs'.

The class contains a single order (Xiphosura), with 1 family.

Class Arachnida

Essentially terrestrial; exoskeleton light to heavy; prosoma wholely or partly covered by carapace; opisthosoma basically of 13 segments plus telson, often reduced; respiratory organs book lungs or tracheae; without compound eyes; abdominal appendages highly modified or absent; consume liquid (usually preliquified animal) food using a pumping pharynx; young stage sometimes with only 3 pairs of legs.

ORDER SCORPIONES

Pedipalps form large chelipeds; prosomal segments fused, covered by carapace; with 7-segmented mesosoma and 5-segmented metasoma, 1st opisthosomal segment absent; telson with sting and poison glands; 2nd opisthosomal segment with pectens; 1 pair of coxal glands, 2 pairs of Malpighian tubules; book lungs; unconcentrated nervous system; terrestrial. 'Scorpions'. 8 families.

ORDER UROPYGI

Prosoma covered by thick carapace; opisthosoma with 12 segments and long whip-like telson; chelicerae fanged; pedipalps large, stout; 1st pair of legs long, thin, used as antennae; acid-secreting opisthosomal glands; 2 pairs of book lungs; both coxal glands and Malpighian tubules; terrestrial. 'Whip scorpions'. 1 family.

ORDER SCHIZOMIDA

Small, weakly sclerotized; without eyes; carapace does not cover last 2 prosomal segments which each have separate dorsal plate; opisthosoma

12-segmented, telson forming small flagellum; pedipalps leg-like, 1st pair of legs antenna-like; 1 pair of book lungs; terrestrial. 2 families.

ORDER AMBLYPYGI

Dorsoventrally flattened; prosoma covered by carapace; telson absent; pedipalps large, raptorial; 1st pair of legs very long, antenna-like; 2 pairs of book lungs; terrestrial. 'Whip spiders'. 3 families.

ORDER PALPIGRADI

Minute; without eyes, Malpighian tubules, circulatory or respiratory systems; last 2 prosomal segments not covered by carapace; chelicerae long, thin; pedipalps leg-like; 1st pair of legs long, used as antennae; telson forming long flagellum; 1 pair of coxal glands; terrestrial. 1 family.

ORDER ARANEAE

Opisthosoma unsegmented, joined to prosoma by stalk; prosoma covered by carapace; chelicerae with poison glands; pedipalps leg-like, forming sperm-transfer organs in male; opisthosomal spinnerets and up to 6 pairs of silk glands; coxal glands and Malpighian tubules; usually, 1 pair of book lungs and 1 pair of tracheae; terrestrial, rarely freshwater. 'Spiders'. 87 families.

ORDER RICINULEI

Heavily sclerotized; hood over mouth and chelicerae; prosoma broadly joined to opisthosoma; prosoma covered by carapace; opisthosoma with 9 fused segments; chelicerae chelate; pedipalps small, leg-like; 3rd pair of legs form copulatory organs in male; anus on posterior tubercle; sieve-tracheae, Malpighian tubules; terrestrial. 1 family.

ORDER PSEUDOSCORPIONES

Small; prosoma covered by carapace, broadly joined to opisthosoma; opisthosoma wide, rounded posteriorly, 11–12 segmented; pedipalps large, forming chelipeds bearing poison glands; coxal glands, tracheae; terrestrial. 22 families.

ORDER SOLPUGIDA

Large; carapace not covering last 2 prosomal segments which have separate dorsal plate articulating with carapace; opisthosoma large, visibly segmented, broadly joined to prosoma; chelicerae massive, forwardly-projecting, chelate; pedipalps leg-like; well-developed tracheal system; both coxal glands and Malpighian tubules; terrestrial. 'Sun spiders'. 12 families.

ORDER OPILIONES

Prosoma and opisthosoma form a single rounded body; legs often very long, tarsus multiarticulate; males with penis, females with ovipositor; coxal glands, tracheae, paired repugnatorial glands; terrestrial. 'Harvestmen', 'Daddylonglegs'. 28 families.

ORDER NOTOSTIGMATA

Prosoma and opisthosoma fused; opisthosoma 13-segmented; mouth parts include a subcapitulum; long legs; 4 pairs of dorsal stigmata; gut with well-marked caeca; pedipalp with claw; terrestrial. 'Leathery mites'. 1 family.

ORDER PARASITIFORMES

Prosoma and opisthosoma fused, covered by carapace, without any demarkating groove, without any discernible segments; mouthparts include a subcapitulum; 1 (rarely 2) pair(s) of ventrolateral stigmata; tracheal system well-developed; coxae not fused to sternum, 1st coxa with opening of coxal glands; terrestrial, free-living and parasitic. 79 families.

ORDER ACARIFORMES

Prosoma and opisthosoma fused, covered by carapace, but demarkated by groove, without any discernible segments; mouthparts include a sub-capitulum; without respiratory organs; coxae fused to ventral body wall; coxal glands discharge through chelicerae; terrestrial and aquatic, free-living and parasitic. 331 families.

Class Pycnogonida

Marine; opisthosoma very small, unsegmented; prosoma divided into a head, with cylindrical proboscis and 3 pairs of appendages (chelicerae, pedipalps, ovigerous legs—non-ambulatory legs used for carrying the eggs), and a segmented trunk of 4–6 segments each with 1 pair of legs borne on the end of large lateral trunk processes; legs often very long (with a span of 75 cm); 2 pairs of eyes on rounded tubercle on posterior head region; lateral gut caeca and parts of gonads extend into legs; without excretory or respiratory organs; haemocoel divided into upper and lower sections by horizontal membrane; nervous system not concentrated; eggs brooded by male and hatch as a protonymphon larval stage with 3 pairs of appendages. 'Sea spiders'.

Pycnogonids may or may not be related to other chelicerates; several authorities regard them as forming a separate phylum. They have been classified in up to 4 orders but no general agreement exists on categories above the level of family, of which there are 10.

Phylum Onychophora

Onychophorans are free-living, bilaterally symmetrical, cylindrically vermiform, terrestrial animals with a head bearing one pair of annular antennae and one pair of claw-like mandibles, and an elongate body with no external segmentation, but with 14–43 pairs of short, unjointed, fleshy, ventral legs ending in paired claws. A thin, flexible, chitinous cuticle covers a body wall composed of epidermal and three muscular layers (circular, oblique and longitudinal); the skeleton is hydrostatic and provided by blood sinuses.

The body surface is covered by tubercles in rings or bands; the gut is a straight tube from the anteroventral mouth, flanked by two oral papillae and surrounded by peribuccal lobes, to the terminal anus; glands opening on the oral papillae secrete a fluid which hardens to sticky, defensive threads; several organs are serially repeated, with one pair in each leg 'segment', including the ostia of the long tubular heart, the excretory metanephridia, and the ganglionic swellings along the paired, ladder-like, ventral nerve cords; no respiratory pigment occurs; anterior nephridia form salivary glands, posterior nephridia gonoducts; the respiratory organs are small tufts of minute, simple tubular tracheae issuing from numerous small spiracles; sense organs include the cephalic antennae, each with a small, simple eye at its base, and sensory bristles on the larger skin tubercles. The sexes are separate, each with elongate paired gonads, the testes producing spermatophores; different species are oviparous, ovoviviparous or viviparous (a placenta attaching the embryo to the uterus wall); oviparous species produce large, yolky eggs which cleave only superficially.

The 70 species of onychophorans occur in a single order with 2 families.

Phylum Tardigrada

Tardigrades are minute (< 1 mm), free-living, bilaterally symmetrical, squat animals inhabiting water films and interstitial spaces. The short cylindrical body bears four pairs of short, unjointed fleshy legs, each ending in 2–11 claws; the head is not demarkated from the trunk and both anterior and posterior ends of the body are bluntly rounded. The body is covered by a mucopolysaccharide and protein cuticle, which may be smooth, ornamented with spines or cirri, or divided into plates; the remainder of the body wall comprises an epidermis only, muscle layers being absent although a lattice of individual, longitudinal, smooth muscle cells occurs in the pseudocoelomic body cavity which forms a hydrostatic skeleton (some accounts suggest that during development five pairs of enterocoelic pouches form, the four anterior pairs subsequently disintegrating and the posterior pair forming the gonadial cavity).

The terminal or subterminal mouth is surrounded by cuticular thickenings, and through it can be protruded a pair of buccal stylets, the muscular pharynx forms a pump, the anus is terminal; three diverticula sometimes occur at the junction of mid and hind gut, these have been regarded as excretory Malpighian tubules; no respiratory or circulatory systems occur; the nervous system includes a bilobed brain, connected by peribuccal cords to a double, ventral nerve cord with ganglia in each leg 'segment'; simple eye spots may be present. The sexes are separate, each containing a single gonad; females are usually by far the more numerous sex, the males of some species being unknown.

Tardigrades or 'water bears' occur in freshwaters, in interstitial marine habitats, and in water films on land. They have been regarded as segmented, coelomate and related to the arthropod phyla, and as unsegmented, pseudocoelomate and related to rotifers and gastrotrichs. The 400 known species are usually divided between 3 orders although the differences between these groups are relatively minor.

ORDER HETEROTARDIGRADA

Cephalic appendages and lateral cephalic cirri; without Malpighian tubules; pharyngeal bulb with rigid cuticular bars, without placoids; legs with simple or complex claws, adhesive discs, or 'toes'; gonopore preanal; marine, freshwater and terrestrial. 5 families.

ORDER MESOTARDIGRADA

Without cephalic appendages; lateral cephalic cirri; Malpighian tubules; pharyngeal bulb with chitinous placoids; legs each with 6–10 simple claws; 4 peribuccal papillae; hot springs. 1 family.

ORDER EUTARDIGRADA

Without cephalic appendages or lateral cephalic cirri; Malpighian tubules; pharyngeal bulb with chitinous placoids; legs each with 2 complex claws; cloaca; freshwater and terrestrial. 2 families.

Phylum Pentastoma

Pentastomans are bilaterally symmetrical, vermiform, annular parasites of vertebrate (mainly reptilian) respiratory passages, with a body comprising a head and annular abdomen. The head may bear five short, lobular processes—an anteroterminal one bearing the jawless mouth, and two leg-like pairs on each side, each bearing a chitinous hook—the protuberances may be lacking, the mouth and the four hooks are then on the body surface; there are no other appendages; the abdominal body wall comprises a porous cuticle, epidermis and layers of circular and longitudinal muscle; a non-coelomic body cavity acts as a hydrostatic skeleton; there is no internal segmentation.

The gut is a straight tube, the pharynx being adapted to pump blood, and the anus being terminal; circulatory, excretory and respiratory systems are lacking; the nervous system is simple, with up to five cephalic ganglia and a ventral cord. The sexes are separate, with internal fertilization; the vagina eventually becomes greatly distended with eggs; the life-cycle involves several larval stages, with unjointed, short, fleshy, hook-bearing legs, and an intermediate vertebrate host. Similar to tardigrades, early embryos have been described as possessing coelomic pouches, which are later lost. Affinities with tardigrades, mites, myriapods and branchiuran crustaceans have all been claimed. 'Tongue worms'.

The 100 species are usually divided between 2 orders.

ORDER CEPHALOBAENIDA

Mouth on anterior margin of pointed head; hooks on fleshy lobes; terminal abdominal annulus with sensory caudal papillae; female genital pore on anterior abdominal annulus; ganglia separate. 2 families.

ORDER POROCEPHALIDA

Ventral mouth on flat head; hooks not on lobes, in horizontal line level with mouth; without sensory caudal papillae; female genital pore on or near posterior abdominal annulus; ganglia fused into single mass. 5 families.

Phylum Uniramia

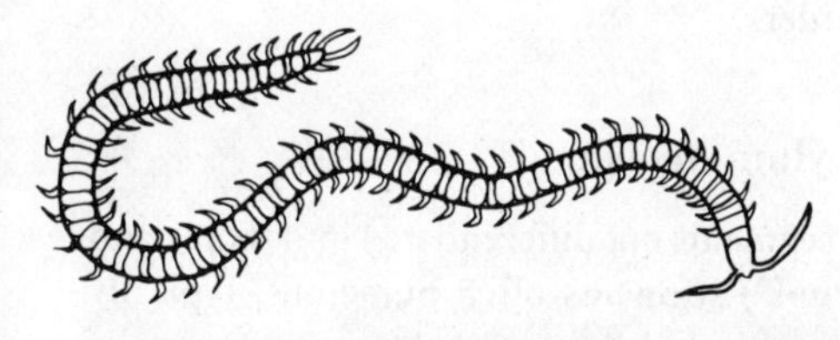

Uniramians are small to medium sized, bilaterally symmetrical and metamerically segmented animals with uniramous jointed limbs and a jointed chitinous exoskeleton which is moulted periodically to permit growth: they were once united with the crustaceans and chelicerates in the phylum Arthropoda, but all three groups are now believed to have had independent origins. The uniramian body comprises a presegmental acron, a postsegmental telson, and from 19 to >200 intervening segments, each bearing one pair of limbs, although the limbs of various segments may be reduced or lost. The head is formed by fusion of the acron and the first four or six segments (dependent on the authority followed); it bears compound eyes and/or ocelli and paired antennae, palpless mandibles, and maxillae (1 or 2 pairs), the maxillae sometimes being fused together. The postcephalic segments either form an elongate, undifferentiated series, or are differentiated into thorax and abdomen.

The head capsule, and sometimes other regions of the exoskeleton, are heavily sclerotized and possess internal skeletal apodemes; the gut possesses both mouth and anus, the mouth being bordered by an upper lip (of epistome and labrum) and a lower lip (of maxillae or labium), and contains the mandibles and a median hypopharynx; no digestive diverticula issue from the mid gut, but from one to many pairs of excretory Malpighian tubules discharge between mid and hind gut; the blood system is open and is not concerned with transport of respiratory gases; blood sinuses form the only body cavity; a dorsal heart is present; the respiratory organs are tracheae, usually opening through lateral spiracles; the nervous system includes a circumoesophageal brain and a double, ventral cord with segmental ganglia subject to various degrees of concentration. The sexes are separate, with gonads varying from one to many and typically with spermatophore production; a distinct larval stage or stages may occur, or the young may hatch as miniature versions of the adult, sometimes with a reduced number of segments and limb pairs.

Uniramians are essentially terrestrial organisms, although several have recolonized freshwaters and a few the margins of the sea.

More than 1 000 000 species are extant, apportioned between 2 subphyla, 10 classes and 50 orders.

Subphylum Myriapoda

Uniramians with postcephalic segments not differentiated into thorax and abdomen; postcephalic (or 'trunk') segments often numerous, typically each with 1 pair of walking legs; mandible with movable endite; coxae of legs have single articulation with sternum.

Class Diplopoda

All but first 4 trunk segments fused into pairs, each 'diplosegment' with 2 pairs of legs, ganglia and heart ostia; simple, 7-jointed antennae; maxillae fused into a gnathochilarium; integument normally reinforced with calcium carbonate; gonopores opening on or near coxae of 2nd pair of legs; 1st trunk segment legless; without compound eyes; each diplosegment covered by a dorsal tergum, 2 lateral pleura, and 2–3 sterna, with various degrees of fusion; young usually hatch with 3 pairs of legs. 'Millipedes'.

ORDER POLYXENIDA

Small; soft, uncalcified integument bearing bristles and trichomes; reduced mouthparts; trunk with 11–13 segments, last 2 segments legless; without gonopods. 4 families.

ORDER GLOMERIDESMIDA

Small; eyeless; trunk of 22 segments; last pair of legs of male forming simple gonopods. 1 family.

ORDER ONISCOMORPHA

Trunk of 12 or 13 segments; last 2 pairs of legs of male forming grasping organs; pygidium and 2nd apparent tergum large, body capable of rolling into tight ball. 'Pill millipedes'. 5 families.

ORDER POLYZONIIDA

Small; leg-like gonopods formed from 2nd pair of legs of 7th segment, 1st pair on that segment normal; male gonopores ending in membraneous penes; pleura and sterna movable; 1st segment large, partly overlapping head. 4 families.

ORDER STEMMIULIDA

1 or 2 pairs of large ocelli; segments laterally compressed, with free sterna and partially fused pleura; long slender legs; anterior pair of legs of 7th segment form gonopods, with long slender flagella, posterior pair of 7th segment greatly reduced; young hatch with many pairs of legs. 1 family.

ORDER SPIROBOLIDA

First 5 leg-bearing segments each with 1 pair of legs; both pairs of legs of 7th segment forming gonopods; 35–60 segments; segments divided transversely into 3 elements. 10 families.

ORDER IULIFORMIDA

Cylindrical body of 30–90 segments; terga divided into 2 elements, sterna fused to tips of tergal arcs, pleura not differentiated; posterior pair of legs of 7th segment reduced to small penes; 1st pair of legs of male highly modified. 19 families.

ORDER TYPHLOGENA

Eyeless; reduced mouthparts; pleura and sterna loosely articulated; terminal segment a complete ring; leg-like gonopods formed by posterior leg-pair of 7th segment and anterior leg-pair of 8th segment. 5 families.

ORDER CHORDEUMATIDA

Last tergal plate with 1–3 pairs of spinnerets; sterna not fused to pleuroterga; posterior leg-pair of 7th segment forms gonopods. 39 families.

ORDER POLYDESMIDA

Eyeless; usually with many segments with paranotal projection on posterior half; 20 trunk segments, with all sclerites fused; anterior leg-pair of 7th segment forming gonopods, posterior leg-pair normal. 27 families.

Class Pauropoda

Small; eyeless; poorly developed mouthparts including 1 pair of maxillae; branched antennae; without heart or limbs on 1st postcephalic segment, mostly without tracheae; gonopores on 3rd trunk segment; trunk usually with 11 segments, the central 9 each with 1 pair of legs; some tergal plates large, extending over 2 segments and bearing long tactile setae; integument soft, uncalcified.

The class contains a single order with 5 families.

Class Chilopoda

Elongate; dorsoventrally flattened; numerous segments, each with 1 pair of legs except 1st trunk segment, which bears 1 pair of large poison fangs, and last 2 segments; simple antennae, 2 pairs of maxillae; integument without calcium carbonate; eyes ocelli, compound or absent; poison fangs occupy undersurface of head; last 2 segments small, forming pregenital and genital segments; female with 1 ovary and 1 gonopore on genital segment, male with 1–24 testes and 2 gonopores on genital segment. 'Centipedes'.

ORDER SCUTIGERIDA

Elongate antennae (up to 400 articles), leg-like 2nd maxillae, 15 leg-bearing segments; legs long, with multiarticulate tarsus; 8 terga, dorsal unpaired spiracles, large compound eyes; eggs laid singly, not brooded; young hatch with 7 pairs of legs. 1 family.

ORDER LITHOBIIDA

Filiform antennae, 15 leg-bearing segments, 2–7 pairs of lateral spiracles; usually with ocelli in lateral clusters; eggs laid singly, not brooded; young hatch with 7 pairs of legs. 5 families.

ORDER SCOLOPENDRIDA

Large head, heavy legs and body, 21 or 23 leg-bearing segments, spiracles on 9 segments; ocelli present or absent; segments deep, terga equisized; eggs laid in clusters, brooded; young hatch with full number of legs. 2 families.

ORDER GEOPHILIDA

Slender 14-articled antennae, long narrow body, 31–> 150 pairs of short legs, spiracles on all segments except first and last; eyeless; eggs laid in clusters, brooded; young hatch with full number of legs; burrow. 13 families.

Class Symphyla

Small; eyeless; trunk with 14 segments, the last fused to telson; first 12 trunk segments each with 1 pair of legs, penultimate segment with cerci and 1 pair of long sensory hairs; dorsal surface with 15–22 tergal plates; antennae long, simple, thread-like; 2nd maxillae fused together; 1 pair of spiracles (on head), tracheae supplying first 3 trunk segments; 1st pair of legs short; young hatch with 6 or 7 pairs of legs; gonopores on 3rd trunk segment; soft uncalcified integument.

The class contains a single order with 2 families.

Subphylum Hexapoda

Postcephalic segments differentiated into 3-segmented thorax, each segment with 1 pair of legs and 2 posterior segments often each with 1 pair of wings, and 11-segmented abdomen without walking legs; abdominal segments subject to reduction or fusion; mandible without movable endite; gonopores terminal or subterminal. 'Insects'.

Class Diplurata

Small; wingless; eyeless; without external genitalia or Malpighian tubules; mouthparts enclosed within pouch; mandible with single articulation, rolling action, protrusible; hypopharynx 3-lobed; abdomen 11-segmented when embryonic, 10th and 11th segments fuse before hatching; gonopores on 9th segment; 7 pairs of abdominal styli, 1 pair of cerci, up to 7 pairs of abdominal spiracles; articles of antennae with muscles.

The class contains a single order (Diplura) with 6 families.

Class Oligoentomata

Small; wingless; without Malpighian tubules, often without tracheae; mouthparts enclosed within pouch; mandible with single articulation, rolling action, protrusible; hypopharynx 3-lobed; abdomen 5-segmented, without spiracles or cerci, with terminal gonopores; antennae 4-articled, first 3 with muscles; 1st abdominal segment with ventral tube (collophore); 3rd abdominal segment with small, partially fused appendages forming retinaculum; appendages of 4th abdominal segment form spring-like furcula operated by haemocoelomic pressure; sometimes with small compound eyes, ocelli vestigeal; eggs cleave totally. Probably neotenous. 'Springtails'.

The class contains a single order (Collembola) with 11 families.

Class Myrientomata

Small; wingless; eyeless; without Malpighian tubules, abdominal spiracles, hypopharynx or cerci; mouthparts enclosed within pouch; stylet-like mandible with single articulation; vestigeal antennae, 1st pair of legs used as antennae; abdomen 11-segmented and with distinct telson; first 3 abdominal segments with small appendages, 1st with stylus; without external genitalia but male gonopores on protrusible phallic complex, gonopores terminal.

The class contains a single order (Protura) with 4 families.

Class Archaeognathata

Wingless; mouthparts ectognathous, mandible with single articulation, rolling action; only basal scape of antenna with muscles; hypopharynx 3-lobed; head with clypeus but without clypeofrontal suture; large maxillary palps; abdomen 11-segmented with styli; tergum of 11th segment prolonged into articulated median caudal filament and flanked by 2 shorter, articulated cercal filaments; head reduced, with ocelli and large compound eyes; female gonopores on 8th segment, male gonopores on 10th; without copulatory organs; with simple egg guide; moulting continues after maturity. 'Bristle-tails'.

The class contains a single order (Microcoryphia) with 1 family.

Class Zygoentomata

Wingless; ectognathous mouthparts, mandible with 2 articulations; hypopharynx with 1 lobe; head with clypeus and clypeofrontal sulcus; small maxillary palps; only basal scape of antenna with muscles; abdomen 11-segmented with 3–8 pairs of styli; tergum of 11th segment prolonged into articulated median caudal filament and flanked by 2 articulated cercal filaments of equivalent length; female gonopores on 8th segment, male gonopores on 10th; without ocelli; reduced compound eyes; without copulatory organs; thoracic and abdominal segments with well-developed transverse tendon system ventrally; tracheal system complex, with longitudinal trunks; moulting continues after maturity. 'Silverfish' and 'firebrats'.

The class contains a single order (Thysanura) with 5 families.

Class Pterygota

One pair of wings (unless lost) on both 2nd and 3rd thoracic segments; ectognathous mouthparts, mandible with 2 articulations; hypopharynx with 1 lobe; maxillary palps small; head with clypeus and clypeofrontal sulcus; only basal scape of antenna with muscles; without abdominal styli except on genital segments; female gonopores on 8th segment, male gonopores on 10th; copulation occurs, female often with ovipositor, eggs with amnion and chorion; epicuticle waterproof; without transverse tendon system in thoracic or abdominal segments; moulting ceases at maturity.

ORDER ODONATA

Minute antennae, very large eyes, 2 pairs of membraneous wings with many small veins; large, often slender, abdomen with accessory male copulatory organs on 2nd segment; nymphs aquatic with posterior tracheal gills and prehensile labium. 'Dragonflies'. 25 families.

ORDER EPHEMEROPTERA

Short-lived adult with small antennae, vestigeal mouthparts, membraneous wings (fore pair the larger), long articulated lateral cerci and a median caudal filament; nymphs aquatic, with paired lateral gills and caudal filaments; subimago with functional wings. 'Mayflies'. 19 families.

ORDER PLECOPTERA

Elongate antennae, reduced mouthparts, membraneous wings (hind pair the larger); usually with long articulated cerci; without ovipositor; nymphs aquatic, bearing long antennae and cerci, with tracheal gills. 'Stoneflies'. 15 families.

ORDER EMBIOPTERA

Females wingless; males with equisized pairs of smoky wings, weakly flying; inhabit silken tunnels, silk secreted by glands in large fore tarsi; 1 or 2 articled cerci, asymmetrical in male, left cercus forming copulatory structure; without ocelli. 'Embiids'. 8 families.

ORDER PHASMIDA

Cylindrical or flattened body with elongate thorax and legs; small or absent, non-functional fore wings; short unsegmented cerci; ovipositor weak; slight metamorphosis. 'Stick-' and 'Leaf-insects'. 11 families.

ORDER ORTHOPTERA

Large prothorax, pronotum extending posteriorly over mesonotum; fore wings usually thickened; hind legs generally large, adapted for jumping; ovipositor large; usually with stridulatory and auditory organs; slight metamorphosis. 'Locusts', 'crickets', etc. 64 families.

ORDER DERMAPTERA

Unjointed cerci forming heavily sclerotized forceps; fore wings short, leathery, without veins; hind wings semicircular, membraneous, or absent; ovipositor reduced or absent; with superlinguae; slight metamorphosis. 'Earwigs'. 11 families.

ORDER GRYLLOBLATTARIA

Without wings or ocelli; compound eyes reduced or absent; long, 8-articled cerci; long filiform antennae; well-developed ovipositor, eversible vesicles on 1st abdominal sternum. 1 family.

ORDER ZORAPTERA

Small; wings, if present, with reduced venation, sheddable; 1-articled cerci; antennae large chain of 9 'beads'; prothorax large; tarsi 2-articled; without ovipositor; 6 Malpighian tubules; slight metamorphosis. 1 family.

ORDER ISOPTERA

Social; polymorphic with several castes; equisized pairs of elongate, membraneous, sheddable wings; antennae short; some with symbiotic intestinal organisms permitting wood-feeding; ovipositor reduced or absent; slight metamorphosis; many individuals with rudimentary or absent genitalia. 'Termites'. 6 families.

ORDER BLATTARIA

Body depressed; pronotum large, with laterally expanded margins, extending over head; thickened fore wings; reduced ovipositor, reduced clypeofrontal sulcus; poorly-developed ocelli; multiarticulate cerci; legs adapted for running. 'Cockroaches'. 5 families.

ORDER MANTODEA

Fore legs raptorial; prothorax elongate; highly mobile head, with large compound eyes, on small neck; thickened fore wings; long antennae; reduced ovipositor; 4 lateral cervical sclerites. 'Mantids'. 8 families.

ORDER PSOCOPTERA

Small; long filiform antennae; prothorax small; ovipositor short; without cerci; reduced, 1–2-articled labial palps; maxillary lacinia rod-like. 'Book lice'. 37 families.

ORDER MALLOPHAGA

Wingless ectoparasites of vertebrates; small antennae in groove beneath head; without ocelli; compound eyes with 2 ommatidia; biting mouthparts; thoracic spiracles ventral; sclerotized lacinia; without cerci and ovipositor; no metamorphosis. 'Biting lice'. 11 families.

ORDER ANOPLURA

Wingless ectoparasites of mammals; without ocelli; compound eyes with 1 ommatidium or absent; piercing/sucking mouthparts retractable into head; thoracic segments fused, with dorsal spiracles; reduced or no lacinal stylets; without cerci and ovipositor, and no metamorphosis. 'Sucking lice'. 15 families.

ORDER THYSANOPTERA

Small; slender-bodied; strap-shaped wings bearing long marginal setae; mouthparts asymmetrical, without right mandible; lacinia slender, pointed; without cerci; 2 pairs of abdominal spiracles; preimaginal instar an inactive, non-feeding pupa. 'Thrips'. 5 families.

ORDER HOMOPTERA

Piercing/sucking mouthparts form a beak, mandibles and laciniae stylet-like enclosed in labium, without labial or maxillary palps; fore wings uniformly hardened, larger than hind wings; usually with ovipositor; without gula; gradual metamorphosis. 'Plant bugs'. 56 families.

ORDER HETEROPTERA

Piercing/sucking mouthparts form a beak, mandibles and laciniae stylet-like enclosed in labium, without labial or maxillary palps; fore wings hardened basally, membraneous at tip, smaller than hind wings; pronotum large; without lateral cervical sclerites; with gula; often without ovipositor; gradual metamorphosis. 'Bugs'. 74 families.

ORDER COLEOPTERA

Sclerotized fore wings forming cover for hind wings when folded; prothorax large, mesothorax small; biting mouthparts; without ovipositor; male genitalia retractable; endopterygote; pupal stage and larva usually with legs. 'Beetles'. 155 families.

ORDER STREPSIPTERA

Endoparasites of other insects, female usually in puparium in host's body; male free-living; male with degenerate mouthparts, large antennae, large metathorax, fore wings reduced to club-shaped balancing organs, and hind wings large and fan-shaped. 8 families.

ORDER HYMENOPTERA

Two pairs of membraneous wings, hind pair the smaller and coupled to fore pair by hooks; reduced metathorax usually fused to 1st abdominal segment forming peduncle; ovipositor modified for stinging, piercing, sawing, etc.; frequently social and polymorphic, with castes, haploid males, parthenogenesis; endopterygote; metamorphosis with pupal and, usually legless, larval stages. 'Ants', 'wasps', 'bees', etc. 90 families.

ORDER RAPHIDIOIDA

Two pairs of similar, membraneous wings; elongate ovipositor; elongate neck membraneous dorsally and covered by anterior extension of pronotum, head elongated posteriorly; endopterygote; metamorphosis, pupa and larva with large sclerotized head and pronotum, and well-developed legs. 'Snake flies'. 2 families.

ORDER NEUROPTERA

Two pairs of similar, membraneous wings; endopterygote; larva with piercing/sucking mouthparts, elongate grooved mandibles covered by elongate laciniae, well-developed legs; Malpighian tubules secrete silk at pupation. 'Lacewings', 'ant-lions', etc. 17 families.

ORDER MEGALOPTERA

Two pairs of similar, membraneous wings; endopterygote; larva with biting mouthparts, aquatic, with segmentally arranged lateral tracheal gills, sometimes with accessory ventral tracheal-gill tufts; adult and larval stipes elongate. 'Alderflies'. 2 families.

ORDER MECOPTERA

Two pairs of similar, membraneous wings; long filiform antennae; long slender legs; elongate abdomen with short cerci; head with ventral rostrum, reduced biting mouthparts; male genitalia prominent; endopterygote; caterpillar-like larva with 3 pairs of short legs, sometimes with abdominal prolegs. 'Scorpion flies'. 8 families.

ORDER DIPTERA

Hind wings reduced to balancing organs, fore wings membraneous; pro- and metathorax small, fused to large mesothorax; mouthparts usually suctorial, forming proboscis with labium expanded distally into fleshy lobes; endopterygote; larva without legs, with reduced head. 'Flies'. 125 families.

ORDER SIPHONAPTERA

Wingless, laterally compressed ectoparasites of vertebrates; short, stout antennae in grooves; vestigeal compound eyes; piercing/sucking mouthparts; without ovipositor; endopterygote; larva elongate, without legs. 'Fleas'. 15 families.

ORDER TRICHOPTERA

Wings hairy, membraneous, fore pair elongate, hind pair broad; mandibles minute or absent; endopterygote; larvae aquatic, often constructing well-camouflaged cases, body with well-developed legs and grasping anal prolegs. 'Caddis flies'. 39 families.

ORDER LEPIDOPTERA

Wings, legs and body covered with broad scales; 2 pairs of membraneous wings, hind pair coupled to fore pair, largely driven by them; mandibles nearly always reduced or absent; maxillae forming long proboscis; endopterygote; larvae with short legs and abdominal prolegs (caterpillars). 'Butterflies' and 'moths'. 138 families.

Phylum Chaetognatha

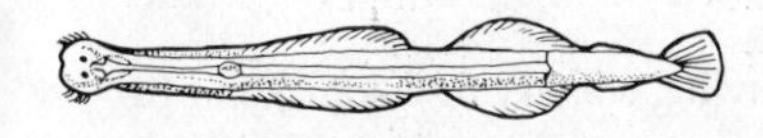

Chaetognaths are free-living, bilaterally symmetrical, unsegmented, vermiform marine animals with a torpedo-shaped body comprising a head, trunk and postanal tail, these regions being demarkated internally by septa and the head being separated from the trunk by a neck. Ventrally, the head bears a large chamber leading into the mouth; flanking the chamber are two rows each with 4–14 large curved movable spines with which prey are grasped, and rows of shorter spines occur in front of the mouth—neither groups of spines are chitinous. A fold of body wall, the hood, can be extended over the head. The trunk and tail bear non-muscular, horizontal fins. A compartmented body cavity occurs; it is enterocoelic in origin but without a peritoneum, and is possibly a secondary cavity in the adult: one compartment occurs in the head, two in the trunk, and one + in the tail.

The body wall comprises a thin, chitinless cuticle, a multilayered epidermis with basement membrane, and paired dorsolateral and ventrolateral bands of longitudinal muscle; the nervous system includes a circumpharyngeal ring enlarged dorsally to form a cerebral ganglion, an anteroventral trunk ganglion, and several lateral ganglia; sense organs include one pair of eyes, each formed by fusion of ocelli, hair fans which detect water-borne vibrations, and a ciliary loop on the dorsal region of head and neck; the gut is a simple tube opening at a ventral anus just anterior to the trunk/tail septum, it includes a muscular pharynx and a pair of intestinal diverticula; no excretory, respiratory or blood systems occur; the reproductive system is hermaphrodite with one pair of ovaries in front of the trunk/tail septum and one pair of testes behind it, self- or cross-fertilization by reciprocal copulation occurs; cleavage of the egg is radial, the deuterostomatous embryo developing directly without a larval stage. 'Arrow-worms'.

The 70 known species are distributed between 2 orders.

ORDER PHRAGMOPHORA

With a transverse ventral musculature; benthic or planktonic. 2 families.

ORDER APHRAGMOPHORA

Without a transverse ventral musculature; planktonic. 3 families.

Phylum Hemichordata

Hemichordates are free-living, solitary or colonial and bilaterally symmetrical, unsegmented benthic marine animals with a body divided into three regions (proboscis, collar and trunk), each with one or two enterocoelic body cavities. The conical or shield-shaped proboscis is connected to the collar by a short stalk; the collar bears the mouth ventrally and may bear dorsally 1–9 pairs of arms each with numerous small ciliated tentacles; in vermiform species the trunk is long and bears the anus terminally; in the tubicolous, colonial species the gut is U-shaped, the anus is borne dorsally on the collar, and the trunk is short and sac-like; the young of some vermiform species have been described as possessing a postanal tail, this may be equivalent to the stalk of tubicolous species. The proboscis bears a single enterocoel and the collar and trunk possess paired cavities, those of the proboscis and collar connect with the exterior via dorsal pores: in many species, however, the peritoneum gives rise to muscle and connective tissue which occludes most of the cavity; these muscles then replace those often associated with the body wall, which remains only as a columnar, ciliated, glandular epithelium.

The gut possesses a long, forwardly-directed diverticulum from the buccal region into the proboscis (for many years considered to be a notochord), and the pharynx usually bears 1–> 100 pairs of U-shaped slits in the dorsal half of its wall, each slit communicates directly with the exterior via a dorsolateral pore and is respiratory, only the ventral part of the pharynx being alimentary. The blood system is partially open, with dorsal and ventral longitudinal vessels, a heart vesicle and a series of sinuses, the blood is colourless. The presumed excretory organ is a glomerulus formed by a series of evaginations of the peritoneum into the proboscis coelom. The nervous system is a basiepithelial plexus, in some areas concentrated into longitudinal cords, e.g., mid-dorsally and mid-ventrally in the trunk, usually with a dorsal invagination of the collar in the form of an often hollow neurochord. The sexes are separate, usually with many pairs of gonads in the trunk coelom; fertilization is external; the eggs cleave radially to form a deuterostomatous ciliated larva or, more rarely, development is direct.

The 100 living species are divided between 3 classes and 4 orders.

Class Enteropneusta

Solitary, freely mobile, vermiform, up to 2.5 m long; without tentacular arms on collar; reduced coelomic cavities; straight gut with terminal anus; many pharyngeal slits, new slits produced throughout life; long proboscis, short collar, very long trunk; with marked powers of regeneration and asexual multiplication; larval stage, when present, a tornaria; inhabit U-shaped mucus-lined burrows, proboscis cilia collecting food. 'Acorn worms'.

The class contains a single order with 4 families.

Class Pterobranchia

Colonial, sessile, inhabiting tubes secreted by shield-shaped proboscis; body sac-like, with stalk and arms on collar; U-shaped gut, anus on collar; at most, 1 pair of pharyngeal slits; without neurochord, mid-dorsal collar only with thickened epidermal plexus; glomerulus poorly developed; asexual multiplication common.

ORDER RHABDOPLEURIDA

Individuals connected by stolons; without pharyngeal slits; 1 pair of tentacular arms; asymmetrical with 1 gonad; colony formed by budding from founding individual. 1 family.

ORDER CEPHALODISCIDA

Individuals in aggregations but not connected; 1 pair of pharyngeal slits; 4–9 pairs of tentacular arms; symmetrical with paired gonads. 2 families.

Class Planctosphaerida

A 'class' erected for a type of giant tornaria larva (up to 25 mm diameter) known from the Atlantic; the spherical body has coelomic cavities, a very sinuous, branched, double band of cilia, and a gut differentiated into pharynx, stomach and intestine. It is assumed to be the larval stage of a hemichordate, but of what type is, as yet, unknown.

Phylum Echinodermata

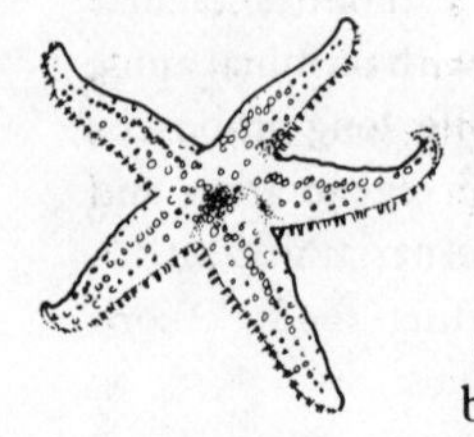

Echinoderms are free-living, mobile or sessile, unsegmented marine animals with a 5-rayed symmetry, in effect radial in some and bilateral in others. A head is lacking, the body possessing a subepidermal system of calcareous spicules or plates, each ossicle being a porous lattice and acting as a single crystal, and a spacious body cavity formed either schizocoelically (in forms with direct development), or enterocoelically (in forms with larval stages) as at least two, and normally six, pouches, one of which forms a water vascular system. This is filled with sea water and comprises a ring canal around the oesophagus, five radial canals, a partially calcified stone canal, often ending at a porous madreporite through which water can be exchanged with the exterior, and a distal series of small outpushings from the canals forming podia, tube feet or tentacles operated hydraulically and arranged along ambulacra. Tube feet may be the locomotory organs, or movement may be achieved through spines or other elements of the calcareous skeleton.

Most echinoderms are expanded in a plane perpendicular to the axis of the gut so that in mobile forms the mouth is in the middle of the lower surface and the anus, if present, is in the middle of the upper surface; in attached, sessile forms the mouth is in the middle of the upper surface. No excretory organs are present, neither is a brain. The nervous and blood systems essentially comprise circumoesophageal rings with five ambulacral elements, the majority of the nervous system being subepidermal; the blood system contains a colourless liquid and is poorly developed, the body cavity forming the main circulatory system; gaseous exchange is effected across the body surface and through papillate extensions of the water vascular system or coelom. The sexes are usually separate, gonads often occurring in multiples of five; fertilization is external and leads to a radially cleaving, deuterostomatous embryo which may develop directly, or via a planktonic larva with long, sinuous tracts of cilia in patterns characteristic of the different echinoderm classes; asexual multiplication is known in some groups.

Echinoderms have a very long fossil history and their classification is based on the fossil record: of the 23 or so known classes, only 6 are extant; these contain 36 orders and 6000 living species.

Class Crinoidea

Attached to substratum for some or all of their life by a stalk from aboral end of cup-like body; mouth directed upwards, surrounded by 5 movable arms which may repeatedly fork; arms with ciliated grooves, numerous side pinnules, finger-shaped tube feet; without madreporite; anus on oral surface; body and arms with heavily calcified, abutting plates and with small tissue volume; separate sexes; gonads on pinnules; eggs brooded or form vitellaria larvae; if stalk lost during life, aboral surface with ring of movable, jointed cirri; suspension feeders. 'Ciroids', 'sea lilies'.

ORDER ISOCRINIDA

Stalked throughout life; with true cirri which adults may use for attachment to substratum. 2 families.

ORDER COMATULIDA

Stalk lost after postmetamorphic juvenile stage; adult mobile, swims with arms or crawls and attaches temporarily using cirri; 5–200 arms. 'Feather stars'. 17 families.

ORDER MILLERICRINIDA

Stalked throughout life; without cirri; body cone-shaped with 5, usually undivided, arms; stalk long, slender, attached by calcareous disc. 1 family.

ORDER BOURGUETICRINIDA

Stalked throughout life; without true cirri; body small, compact, not sharply demarkated from stalk; attached to hard substrata by irregular plate, to soft substrata by branching rootlets. 3 families.

ORDER CYRTOCRINIDA

Stalked throughout life; stalk short, massive, usually without separate ossicles, sometimes missing and body attached directly to substratum; without cirri; 10 arms. 1 family.

Class Somasteroidea

Very flat, freely mobile; broad petaloid arms, small central disc; arms with very long pinnules, without permanent ambulacral grooves, with small simple sucker-less tube feet, with ampullae of tube feet in external cavities; disc with mouth in centre of lower surface, with blind-ending gut, with marginal aboral madreporite.

Somasteroids gave rise to asteroids and ophiuroids in the lower Ordovician; only one species, *Platasterias latiradiata*, of the order Goniactinida, remains extant.

Class Asteroidea

Flat, freely mobile; with central disc from which radiate 5+ broad tapering arms, arms and disc not sharply demarkated; mouth in centre of lower surface, anus, if present, at centre of upper surface; permanent ambulacral groove along each arm with margin of paired, unfused ambulacral ossicles; tube feet with internal ampullae, often with suckers; calcareous skeleton diffuse, body cavity extensive; gaseous exchange via tube feet and coelomic papulae; with aboral madreporite; with pedicellariae; some species hermaphrodite; with direct development or bipinnaria and brachiolaria larvae. 'Starfish'.

ORDER PLATYASTERIDA

With simple unstalked pedicellariae; arms bordered by 1 row of large plates; tube feet in 2 rows, with double ampullae, without suckers; without anus or intestine. 1 family.

ORDER PAXILLOSIDA

With simple unstalked pedicellariae; arms bordered by 2 rows of plates; tube feet large, pointed, in 2 rows, mostly without suckers; anus and intestine sometimes present. 5 families.

ORDER VALVATIDA

If present, pedicellariae pincer-like; arms bordered by 2 rows of plates; tube feet with suckers, with double ampullae, in 1–2 rows; body rigid; anus present. 9 families.

ORDER SPINULOSIDA

With 5–18 arms bordered by 2 inconspicuous rows of small plates; pedicellariae rarely present, if so, only as small groups of spines; tube feet in 2 or 4 rows, with suckers, with single or double ampullae; anus present. 12 families.

ORDER FORCIPULATIDA

With 5–50 arms, not bordered by rows of plates; with complex, pincer-like pedicellariae; tube feet with suckers, with double ampullae, in rows of 4 (or 6); arms circular in section. 3 families.

ORDER EUCLASTERIDA

With 9–15 arms, often very long, not bordered by rows of plates, sharply demarkated from small disc; pedicellariae all with crossed valves; tube feet long, with suckers, with single ampullae, in 2 rows; deep sea. 1 family.

Class Ophiuroidea

Very flat, freely mobile; with narrow arms sharply demarkated from central disc; mouth in centre of lower surface, without intestine or anus; well-developed calcareous skeleton occupying most of body volume, body cavity correspondingly reduced; ossicles of arms fused into longitudinal series of 'vertebrae', ambulacral grooves completely enclosed; tube feet without suckers, used in feeding not locomotion, with internal ampullae; arms used in movement; gonads discharge through paired invaginations of body wall, bursae, opening via slits at base of arms, bursae also sometimes used as brood chambers and for gaseous exchange; madreporite on oral surface; without pedicellariae; larval stage an ophiopluteus. 'Brittlestars', 'basketstars'.

ORDER OEGOPHIURIDA

Ambulacral grooves enclosed only by skin; without dorsal and ventral arm plates; without bursae or oral shields; gastric caeca extend into arms; madreporite marginal. 1 family.

ORDER PHRYNOPHIURIDA

Ambulacral grooves enclosed by plates; arms without dorsal plates, with small ventral plates, with lateral plates small and ventral, arms sometimes much branched, coilable; disc and arms naked or covered by skin; with oral shields; gastric caeca not extending into arms. 'Serpentstars', 'basketstars'. 5 families.

ORDER OPHIURIDA

Ambulacral grooves enclosed by plates; arms incapable of much coiling, with dorsal, lateral and ventral plates, arms unbranched; madreporite on oral shield; gastric caeca not extending into arms. 11 families.

Class Echinoidea

Freely mobile; spherical or flattened, armless body; calcareous skeleton well developed, large plates attached to each other to form rigid test, each ambulacral and interambulacral area with 2 rows of plates to total 20 vertical columns; mouth in middle of lower surface, sometimes displaced, usually surrounded by chewing apparatus of 5 teeth (Aristotle's lantern); anus in middle of aboral surface or displaced 'posteriorly' in species with secondary bilateral symmetry; test with strong, movable spines and stalked pedicellariae; each ambulacral area with 2 rows of locomotory tube feet; larval stage an echinopluteus or development direct. 'Sea urchins'.

ORDER CIDAROIDA

Ambulacral plates each with 1 tube foot; wide interambulacral areas, each interambulacral plate with 1 massive, pencil-like spine surrounded by small scrobicular spines; teeth of Aristotle's lantern not keeled; without gill slits. 2 families.

ORDER ECHINOTHUROIDA

Test flexible, with poison spines; ambulacral and interambulacral plates extend across peristomial membrane; long delicate club-shaped oral spines; development direct; deep sea. 1 family.

ORDER DIADEMATOIDA

With hollow spines; 10 buccal plates in circle on peristomial membrane; prominent gill notches at margins of peristomial membrane. 4 families.

ORDER PEDINOIDA

With solid spines; perforate primary tubercles; without keels on teeth of Aristotle's lantern; 10 buccal plates in circle on peristomial membrane; gill notches slight. 1 family.

ORDER SALENOIDA

With solid spines; imperforate tubercles; 10 buccal plates on peristomial membrane; each interambulacral plate with 1 long slender spine; anus displaced off-centre by large suranal plate; small; deep sea. 1 family.

ORDER PHYMOSOMATOIDA

Aristotle's lantern stirodont; some ambulacral plates fused; 10 buccal plates on peristomial membrane; without perforate tubercles on suranal plate. 2 families.

ORDER ARBACIOIDA

Aristotle's lantern stirodont; naked areas of test aborally and periproct with 4–5 large conspicuous plates covering anus; without secondary spines; large oral membrane with 10 buccal plates. 1 family.

ORDER TEMNOPLEUROIDA

Aristotle's lantern camarodont; usually sculptured test with solid spines; 10 buccal plates on peristomial membrane. 2 families.

ORDER ECHINOIDA

Aristotle's lantern camarodont; 3–16 pairs of pores per ambulacral plate; tubercles imperforate; 10 buccal plates on peristomial membrane; without sculpturing of test. 4 families.

ORDER HOLECTYPOIDA

Irregular; slightly flattened test; mouth mid-oral; anus displaced 'posteriorly', sometimes lying near mouth; Aristotle's lantern stirodont, with keeled teeth; without petaloids. 1 family.

ORDER CLYPEASTEROIDA

Irregular; flattened or greatly flattened test; mouth small, mid-oral; anus 'posterior'; stirodont Aristotle's lantern; aboral ambulacra petaloid; without oral phyllodes. 'Sand dollars'. 9 families.

ORDER CASSIDULOIDA

Irregular; without Aristotle's lantern; test round or oval, flattened; mouth mid-oral; anus 'posterior'; poorly-developed petaloid aboral ambulacra; oral phyllodes; interambulacra at edge of mouth form raised bourrelets; live partially buried. 4 families.

ORDER SPATANGOIDA

Irregular, bilaterally symmetrical; test oval to elongate; without Aristotle's lantern; mouth 'anterior'; anus 'posterior'; anterior aboral ambulacra not petaloid, often sunken, other aboral ambulacra petaloid; oral phyllodes; tube feet and spines regionally specialized. 'Heart urchins'. 10 families.

ORDER NEOLAMPADOIDA

Irregular, small, oval; without Aristotle's lantern; mouth mid-oral; anus 'posterior'; without petaloid aboral ambulacra; without oral phyllodes. 1 family.

ORDER HOLASTEROIDA

Irregular; oval, elongate, bell- or bottle-shaped; without Aristotle's lantern; thin fragile test; petaloids poorly developed or absent; without phyllodes; 3 anterior aboral ambulacra meet separately from 2 posterior ambulacra; deep sea. 4 families.

Class Holothuroidea

Freely mobile; without arms; body elongate, mouth at one end, anus at other; bilaterally symmetrical, 3 ambulacra in contact with substratum forming 'ventral' sole, remaining 2 ambulacra 'dorsal'; calcareous skeleton reduced to separate microscopic ossicles; water vascular system with ring of 8–30 finger-like or branched tentacles around mouth; body wall leathery, 5 longitudinal muscles, circular muscle; spacious body cavity; 1 or 2 gonads; 2 respiratory trees often open into rectum, sea water pumped into them by cloaca; partial evisceration or discharge of Cuvierian tubules used to deter predators; larval stage, if present, an auricularia, doliolaria or barrel-shaped vitellaria. 'Sea cucumbers'.

ORDER DENDROCHIROTIDA

With 10–30 highly branched, retractable tentacles; respiratory trees present; madreporite free in body cavity; 2 gonads; tube feet present; epi- or infaunally benthic. 7 families.

ORDER DACTYLOCHIROTIDA

With 8–30 finger-like, retractable tentacles; rigid U-shaped body enclosed in test of imbricating plates; tube feet present; infaunal in deep sea. 3 families.

ORDER ASPIDOCHIROTIDA

With 15–30 shield-shaped tentacles; respiratory trees present; dorsal tube feet form papillae, ventral tube feet locomotory; usually epifaunal, shallow water. 3 families.

ORDER ELASIPODA

Fragile, gelatinous, with shield-shaped tentacles; without respiratory trees; dorsal tube feet often form elongate papillae or long processes; pelagic or benthic, deep sea. 5 families.

ORDER APODIDA

Elongate, vermiform; with 10–25 finger-like or pinnate tentacles; without respiratory trees or tube feet; thin transparent body wall often with anchor-shaped ossicles; infaunal in soft sediments. 3 families.

ORDER MOLPADIIDA

Stout, posterior region drawn out into distinct tail; with 15 finger-like tentacles; without tube feet except as anal papillae; respiratory trees present; mostly infaunal in soft sediments. 4 families.

Phylum Chordata

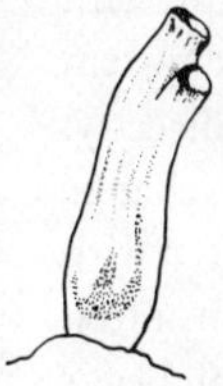

Chordates are free-living, bilaterally symmetrical animals with an internal skeletal rod, the notochord, extending along the dorsal midline, with a hollow dorsal nerve cord running dorsal to the notochord, with a postanal tail containing extensions of the nerve cord and notochord, and with a pharynx perforated by slits leading from the lumen of the gut through to the outside of the body, water being induced to enter the gut through the mouth and to pass out through these slits. The notochord is often replaced by cartilaginous or bony material to form an endoskeleton of many elements, some of them forming protective casings around the dorsal central nervous system; the pharyngeal slits serve a feeding and/or respiratory function in aquatic chordates, they are greatly reduced in terrestrial forms.

The different subphyla of chordates differ markedly in their other anatomical features (e.g., whether they are segmented, possess a body cavity, bear appendages or jaws, have a differentiated head, etc.) and so these will be described separately for each subphylum. Some 40 000 chordate species are extant.

Subphylum Urochordata

Sessile or free-swimming, solitary or colonial, marine; body unsegmented, mostly occupied by relatively enormous pharynx perforated by numerous slits (stigmata); without body cavity; adults without notochord, postanal tail or hollow dorsal nerve cord, although these may occur in the larva; body encased in external fibrous test or tunic, usually containing cellulose. Pharynx forms filter-feeding system, water entering through branchial siphon by ciliary action, passing through stigmata, particles being trapped in mucus secreted by longitudinal ventral endostyle, and filtered water passing into cavity surrounding pharynx, atrium, before leaving body through single atrial siphon. Remainder of gut small, often U-shaped, with anus near atrial siphon.

Body wall with two epidermal layers separated by mesenchyme containing muscle bands, outer epidermis secretes tunic (although external, blood vessels may extend into tunic); blood system with sinuses in mesenchyme, tubular heart with periodically reversing direction of pumping, and with high concentrations of vanadium in some blood cells; nervous system simple, ganglion between siphons from which nerves issue directly; without excretory organs or internal skeletal elements; gaseous exchange across walls of pharynx; usually simultaneously hermaphrodite, often with one ovary and testis; asexual multiplication by budding common, often resulting in formation of colonies, some enclosed in common tunic; eggs brooded or develop externally; embryo deuterostomatous, in sessile species forming tadpole larva with muscular postanal tail with notochord and hollow dorsal nerve cord; tail resorbed when larva attaches to substratum and metamorphoses. 'Tunicates'.

The Urochordata includes 3 classes and 8 orders.

Class Ascidiacea

Sessile; branchial and atrial siphons close together; life-cycle simple, without alternation of sexual and asexual generations. 'Sea squirts'.

ORDER APLOUSOBRANCHIA

Body divided into thorax and abdomen or thorax, abdomen and postabdomen; gonads in intestinal loop; pharynx wall not folded, with <20 rows of stigmata; form compound colonies. 3 families.

ORDER PHLEBOBRANCHIA

Body never with postabdomen; internal longitudinal bars on pharyngeal wall; gonads in intestinal loop; eggs not brooded, tadpole larva simple; mostly solitary, often large. 7 families.

ORDER PLEUROGONA

Body not divided into regions; internal longitudinal bars on pharyngeal wall, with internal longitudinal folding of pharyngeal wall; gonads alongside pharynx, arising from atrial wall; mostly solitary. 3 families.

ORDER ASPIRACULATA

Pharynx reduced, without stigmata; branchial siphon forming prey-capture organ with 6 prehensile lobes; solitary; deep sea. 1 family.

Class Thaliacea

Planktonic; branchial and atrial siphons at opposite ends of barrel-shaped or fusiform body; exhalant water current provides propulsive force for movement; either colonial or alternate between solitary and colonial forms; asexual buds formed on ventral stolon.

ORDER PYROSOMIDA

Colonial; zooids in common cylindrical test open at one end only; zooids with branchial siphons opening on external colony surface, atrial siphons on internal surface; pharynx with numerous stigmata; without free-swimming larva; colonies up to 2 m long. 'Pyrosomes'. 1 family.

ORDER DOLIOLIDA

Solitary generation reproduces sexually to form asexually-multiplying colonial generation; polymorphic individuals; tunic thin, gelatinous, transparent, with 8–9 complete hoops of muscle encircling body; pharynx with few to many stigmata; tadpole larval stage. 'Doliolids'. 1 family.

ORDER SALPIDA

Solitary generation multiplies asexually, chains separating to yield solitary sexual generation; tunic thick, gelatinous, with incomplete hoops of muscle encircling body; pharynx with 2 large stigmata; viviparous, without larval stage. 'Salps'. 1 family.

Class Larvacea

Small, solitary, neotenous; with large tail; secrete a gelatinous 'house', through filters of which water is drawn by beating movements of the tail; when the 'house' is clogged it is discarded and a new one secreted; pharynx with 2 stigmata opening directly on to body surface; endostyle reduced; tail with notochord and nerve cord, much larger than body.

The class contains the single order, Copelata, with 3 families.

Subphylum Cephalochordata

Free-swimming, solitary, benthic marine; laterally compressed, fish-like body, tapering at each end; persistent notochord extending whole length of body, but with no other skeletal elements; hollow dorsal nerve cord extending almost whole length of body, not dilated anteriorly to form brain; small postanal tail; segmentally-arranged myotomal muscle blocks; enterocoelic body cavity, which in the adult is small; large pharynx perforated by numerous stigmata, surrounded by extensive atrial chamber with a single atriopore, and with longitudinal ventral endostyle. Pharynx forms filter-feeding system, water entering through anteroventral mouth, passing through stigmata, particles being trapped in mucus from the endostyle, and filtered water leaving atrium through atriopore; cilia propel feeding water-currents.

Body covered by very thin, monolayered epidermis; anterior end with buccal cirri, wheel organ and velar tentacles, but without differentiated head; blood system includes vessels and sinuses, the blood being colourless; excretory organs resemble protonephridia but formed by peritoneal cells; muscle fibres send processes to dorsal nerve cord, peripheral fibres without myelin sheath; body with continuous dorsal, caudal and posterioventral fins. Sexes separate; numerous gonads in paired rows or only on right side, segmentally arranged; external fertilization and deuterostomatous development. 'Lancelets'.

The subphylum contains a single class and order with 2 familes.

Subphylum Vertebrata

Free-moving, solitary, aquatic and terrestrial; with few pharyngeal slits, not associated with mucociliary filter-feeding, endostyle forming thyroid gland; endoskeleton with notochord usually enclosed in additional material (cartilage or bone) and divided into linear series of vertebrae, with protective cranium around dilated anterior region of dorsal nerve cord, with nerve cord enclosed within longitudinal skeletal element; a distinct head region with paired eyes, olfactory and other sense organs, and brain; body cavity formed enterocoelically or by a variety of secondary methods; segmentally arranged nerves and muscles; closed blood system with well-developed heart; excretory kidneys operate by ultrafiltration; sexes usually separate. Body covered by skin with dermal and epidermal layers, often with protective dermal plates, bones or scales, or with epidermal scales or scale-derivatives; in all but one group, jaws and two pairs of lateral appendages present; gaseous exchange effected by gills in pharyngeal slits or lungs associated with the gut.

The vertebrates are the best known and most intensively studied of organisms, and as a result are 'over-classified' in relation to other groups of animals; many vertebrate 'orders' are equivalent to the families of most groups of invertebrates (especially true of the birds), and some 'classes' are more akin to invertebrate orders. Hence, to maintain a degree of equivalence of treatment within this synoptic account, the level of superorder will often be the lowest taxon into which vertebrate classes will be subdivided, this can be regarded as roughly equivalent to order as used elsewhere in the book. On this basis, the treatment here recognizes the 7 'classes' into which it is customary to divide the vertebrates and some 40 smaller groupings: some texts list over 100 'orders' of living vertebrates.

Class Agnatha

Fish-like, aquatic; without jaws; without paired lateral appendages; well-developed, persistent notochord with irregular cartilaginous thickenings; pharynx with series of somatopleural cartilages supporting the sac- or pouch-like gills; scaleless, boneless, eel-like; with up to 16 pairs of gills; ear with 1 or 2 semi-circular canals; larval stage of one group with endostyle and pharynx perforated by numerous stigmata, functioning as filter-feeding system. 'Lampreys' and 'hagfish'.

The class contains a single superorder, Cyclostomata, with 2 families (each sometimes placed in a separate superorder).

Class Chondrichthyes

Medium to large sized, fish-like, aquatic; with jaws bearing teeth; paired pectoral and pelvic fins with internal skeletal supports; notochord often replaced by cartilaginous skeleton with local calcification; pelvic fins of male with copulatory claspers; skin with dermal/epidermal placoid scales; skull without sutures between different elements, palatoquadrate cartilage forming upper jaw; intestine with spiral valve; with ventral nostrils; without lungs or swim bladder; few large eggs produced, protected in individual horny or leathery cases, or within female; mostly marine.

Superorder Neoselachii

Upper jaw not fused to skull, supported by hyomandibular cartilage; mouth anteroventral; 5–7 gill slits opening separately to exterior; rigid dorsal fins; vertebral centra partly or completely calcified; spiracle present. 'Sharks', 'rays', etc. 47 families.

Superorder Holocephalii

Upper jaw fused to skull, without hyomandibular cartilage; mouth terminal; 4 gill slits covered by operculum; anterior dorsal fin erectile; notochord not constricted segmentally, surrounded by calcified rings, without vertebral centra; without spiracle; frontal clasper on head; long whip-like tail. 3 families.

Class Osteichthyes

Small to large sized, fish-like, aquatic; with jaws usually bearing teeth fused to bone supporting them; skeleton of bone usually replacing notochord (sometimes secondarily reduced to cartilage), bony skull with sutures between different elements, upper jaw formed of premaxillary and maxillary bones; paired pectoral and pelvic fins supported by segmented dermal fin rays; swim bladder or lung (sometimes lost) present; skin with bony scales, often reduced in size and/or thickness; gill slits covered externally by operculum; fertilization usually external.

Superorder Dipnoi

Large; diphycercal tail; single or paired swim bladder functions as lung; paired fins with central longitudinal skeletal axis, broad, flipper-like or filamentous; without premaxillae or maxillae; freshwater. 'Lungfish'. 3 families.

Superorder Coelacanthini

Large; diphycercal tail; swim bladder slender, fat-filled; spiral valve in gut; massive, unconstricted notochord, without vertebral centra; scales cosmoid; paired fins with central longitudinal skeletal axis; skull with transverse joint; marine. 'Coelacanth'. 1 family.

Superorder Polypterini

Elongate, cylindrical; large ganoid scales; spiral valve in intestine; 2-chambered swim bladder; spiracle present; without pelvic fins; fan-like pectoral fins with fleshy stalk; freshwater. 'Bichirs'. 1 family.

Superorder Acipenserini

Large; skeleton mostly cartilaginous; skin scaleless or with few large bony scutes; heterocercal tail; spiral valve in intestine; unconstricted notochord; weak suctorial mouth; spiracles present; swim bladder large; mostly freshwater. 'Sturgeons' and 'paddlefish'. 2 families.

superorder Semionotini

Elongate, with elongate head; large, ganoid scales, without dentine; heterocercal tail; swim bladder functions as lung; simple spiral valve in intestine; vertebrae opisthocoelous; mainly freshwater. 'Gar-pike'. 1 family.

Superorder amiimorpha

Large; round cycloid scales; tail heterocercal but with homocercal external appearance; maxilla reduced; with long dorsal fin, 2 gular plates, tubular nostrils, well-developed teeth; freshwater. 'Bowfin'. 1 family.

Superorder Elopomorpha

With leptocephalus larva; teleost; adults often eel-like; ethmoidal commissure enclosed in bone, associated with small rostral ossicles; branchiostegals numerous; marine. 32 families.

Superorder Clupeomorpha

Silvery, laterally compressed; teleost; diverticula of swim bladder forming bullae within ear capsule; few lateral-line pores on trunk; skull with temporal foramina, pre-epiotic fossae, auditory fenestrae, recessus lateralis; mainly marine. 4 families.

Superorder Osteoglossomorpha

Teleost; with paired sub-branchial tendon bones at base of 2nd gill arch; primary bite between parasphenoid and basi- and glosso-hyals; freshwater. 6 families.

SUPERORDER PROTACANTHOPTERYGII

With fleshy, dorsal, adipose fin, usually ventral pectorals, abdominal pelvics; with trends towards exclusion of maxillae from gape, development of premaxillary process, loss of mesocoracoid; basal teleost group from which spiny-rayed forms evolved. 51 families.

SUPERORDER OSTARIOPHYSI

Teleosts with series of 3 or 4 movable bony ossicles linking swim bladder to ear; ventral pectorals, abdominal pelvics; hypurals of tail on 1 centrum; upper jaw often protrusible; mainly freshwater. 57 families.

SUPERORDER PARACANTHOPTERYGII

Teleosts with anterior pelvics, lateral pectorals, non-protractile upper jaw, reduced or absent pleural ribs; caudal skeleton, when present, of 2 large hypurals (sometimes fused) on separate centra; with up to 13 radials in pectorals; carnivorous; mainly marine. 30 families.

SUPERORDER ATHERINOMORPHA

Small, surface feeding teleosts with 4 radials in pectorals recessed into excavation; upper jaw often protractile, without ascending processes or palatopremaxillary or ethmomaxillary ligaments; caudal skeleton with 2 large hypural plates on terminal half-centrum, never more than 4 hypurals. 16 families.

SUPERORDER ACANTHOPTERYGII

Teleosts with anterior pelvics, lateral pectorals, well-developed pleural ribs, caudal skeleton of >6 hypurals on 2 centra or all hypurals on 1 centrum; pectorals with up to 4 radials; often with protractile upper jaw, fin spines and ctenoid scales; mainly marine. 217 families.

Class Amphibia

With paired lateral appendages forming jointed legs (occasionally lost); aquatic larval stage respiring through gills or skin, with longitudinal fins, often with lateral-line system; adult aquatic, amphibious or terrestrial, with reduced pharyngeal clefts without respiratory function, gaseous exchange across lungs; with bony skeleton including well-developed skull and jaws; skin glandular without epidermal scales, hair, bony plates, etc.; eggs anamniotic, laid in wet conditions; with 1 sacral vertebra, 2 occipital condyles; hyomandibular forms stapes; neoteny common. 'Amphibians'.

ORDER ANURA

Tail-less; hind legs modified for jumping, hind foot with elongate astragalus and calcaneum; 5–9 presacral vertebrae; larva without true teeth, with tail; tail resorbed during metamorphosis; never neotenous. 'Frogs' and 'toads'. 23 families.

ORDER URODELA

With long tails; limbs equisized, sometimes reduced; skull partly cartilaginous, with frontal and parietal bones unfused; larvae with teeth and persistent external gills; often neotenous. 'Newts', 'salamanders', etc. 8 families.

ORDER GYMNOPHIONA

Elongate, leg-less, annulate, often tail-less, subterranean; with small or no eyes, fused skull bones, numerous vertebrae, rudimentary left lung, penis; a sensory cephalic tentacle. 'Caecilians'. 5 families.

Class Reptilia

With paired lateral appendages forming jointed legs (occasionally lost); without larval stage; adults terrestrial or secondarily aquatic, with greatly reduced pharyngeal slits, never having respiratory function, gaseous exchange across lungs; with bony skeleton including well-developed skull and jaws; skin with covering of ectodermal scales; eggs amniotic, shelled, in some retained within body of female; internal fertilization, with penis; 2+ sacral vertebrae, 1 occipital condyle. 'Reptiles'.

ORDER CHELONIA

Body covered by bony shell (dorsal carapace and ventral plastron) to which vertebrae and ribs often fuse, only head/neck, tail and distal limb elements protrude from shell; with tooth-less beak; penis single; without fenestrae in dermal skull bones. 'Tortoises' and 'turtles'. 11 families.

ORDER RHYNCHOCEPHALIA

Lizard-like, with 4 limbs; 2 fenestrae in temporal region of skull, 2 temporal arches; teeth fused to jaw; without copulatory organs; with gastralia. 'Tuatara'. 1 family.

ORDER SQUAMATA

With 4 legs or commonly legs greatly reduced or absent; 2 fenestrae in temporal region of skull, no or 1 temporal arch; without gastralia; male with paired hemipenes; quadrate frequently movable; teeth not fused to jaw. 'Lizards', 'snakes', 'amphisbaenians'. 33 families.

ORDER CROCODILIA

Large, elongate, amphibious; large tail; shorter fore legs; 2 temporal and other smaller fenestrae in skull; with secondary palate, simple thecodont teeth, procoelous vertebrae, 4-chambered heart; single penis; pubis excluded from perforate acetabulum. 'Crocodiles'. 1 family.

Class Aves

With paired lateral appendages, hind pair forming jointed legs, fore pair forming jointed wings (occasionally reduced or forming flippers); body and wings covered by ectodermal feathers, distal leg elements covered by scales; head with tooth-less beak, large eyes and brain, movable upper jaw; body short, with large keeled sternum, pelvic girdle fused to fused sacral vertebrae; tail minute, but tail feathers often long, caudal vertebrae fused to form short pygostyle; bony skeleton and skull well developed, light, several bones with air spaces; eggs amniotic, shelled, often brooded; no larval stage; pharyngeal slits greatly reduced, never with respiratory function, lungs organs of gaseous exchange; heart 4-chambered, without left aortic arch; lungs small, inflexible, extend into complex air sacs; can sustain homoiothermy endothermally; essentially aerial, feathered, thecodont reptiles. 'Birds'.

Superorder Palaeognathae

Large, mainly flightless; without keel on sternum; with massive bones, loose, plumose feathers; nasal bone not fused to maxillary (permitting rhynchokinesis); pterygoid meets large vomer in palate. 'Ostrich', etc. 6 families.

Superorder Impennes

Flightless, with fore limbs forming flippers; scale-like feathers uniformly over body; legs short; sternum keeled, 'flight' muscles large; no apteria; tarsometatarsus incompletely fused; palatine separates pterygoid and small vomer; coastal marine. 'Penguins'. 1 family.

Superorder Neognathae

Flying or secondarily flightless, with keeled sternum, large flight muscles, wings with primary and secondary feathers, tail with rectrices attached to pygostyle; tarsometatarsus fused; palatine separates pterygoid and small vomer. 165 families.

Class Mammalia

With paired lateral appendages forming jointed legs, fore pair sometimes modified as flippers or wings, hind pair sometimes absent; whole or part of body covered by ectodermal hairs, sometimes matted together to form spines, scales or other structures; bony skeleton and skull well developed, lower jaw formed by single bone, quadrate and articular bones forming ear ossicles (incus and malleus), dermal elements of pectoral girdle reduced or absent, false palate present; jaws usually with teeth of more than one type, teeth replaced only once; no larval stage; pharyngeal slits greatly reduced, never with respiratory function, lungs organs of gaseous exchange; heart 4-chambered, without right aortic arch; embryos amniote, either with shell or retained within uterus, usually viviparous; young nourished by secretions from mammary glands of females; muscular diaphragm divides thoracic and abdominal cavities; erythrocytes without nuclei; can sustain homoiothermy endothermally; essentially synapsid reptiles with single dentary/squamosal jaw articulation. 'Mammals'.

Superorder Ornithodelphia

Oviparous; with nippleless mammary glands, cloaca, dermal pectoral bones, unfused uteri, epipubic bones, cervical ribs; without auditory bullae; without teeth in adults; males with poison-gland-bearing horny spurs on ankles. 'Platypus' and 'echidna'. 2 families.

Superorder Marsupialia

Viviparous, without chorioallantoic placenta with villae; young born after brief gestation, suckled and develop further in abdominal marsupium; with epipubic bones, scrotum anterior to penis, separate vaginae and uteri; penis bifid; without corpus callosum in brain. 'Marsupials'. 16 families.

EUTHERIAN ORDERS

Viviparous, with chorioallantoic placenta with villae; young born at advanced stage of development; without epipubic bones; single vagina, simple penis; brain with corpus callosum; middle ears covered by petrosal bone.

Modern classifications list some 16–22 extant 'orders' of living placental mammals (together with a further 20 extinct 'orders'), with 114–125 families. Several schemes for apportioning these 'orders' into cohorts, superorders, etc. are current, but none has achieved general adoption: the following 8 groupings are broadly recognized.

Edentata. Without enamel on teeth; teeth simple, peg-like; low body temperature; extra articulatory surfaces on trunk vertebrae; testes with partial or no descent; sometimes with dermal ossifications; terrestrial. 'Sloths', 'armadillos', etc. 3 families.

Pholidota. Body covered by large, imbricating scales; skull smooth, conical, without teeth, with incomplete zygomatic arch, with slender dentary; gut without caecum; terrestrial. 'Pangolins'. 1 family.

Anagalida. Large hind legs adapted for jumping and largely bipedal progression; gut with large caecum; jaw with high condyle; all ancestral molar teeth retained; cheek teeth often prismatic; terrestrial. 'Rabbits', 'elephant shrews', etc. 3 families.

Insectivora. With long olfactory capsules; maxillary bone extending into orbit wall; without gut caecum or postorbital bar; reduced jugal and pubic symphysis; often small; terrestrial and freshwater. 'Shrews', 'hedgehogs', etc. 7 families.

Carnivora. With well-developed canine and carnassial teeth; scaphoid, lunar and centrale carpal bones fused; often large; predatory; terrestrial and aquatic. 'Bears', 'cats', 'seals', etc. 12 families.

Rodentia. Incisors with enamel only on anterior surface, with persistently open roots; long diastema; generally small; herbivorous; terrestrial and freshwater. 'Mice', 'squirrels', etc. 33 families.

Archonta. A collection of eutherians characterized mainly by lacking specializations possessed by other groups; with characteristic ankle joint; some with fore limbs forming wings ('bats'), others with large brains ('monkeys', 'apes' and 'man'). 32 families.

Ungulata. Mainly herbivorous, with high-crowned grinding teeth and long legs moving in fore-and-aft plane, distal limb bones elongated, digits often reduced in number, stand and move on toes; some lines aquatic with dorsoventrally flattened tail, paddle-like fore limbs, without hind limbs; usually large to very large. 'Elephants', 'deer', 'horses', 'whales', 'manatees', etc. 25 families.

References and Further Reading

General

Margulis L. & Schwartz K. V. (1982) *Five Kingdoms: An Illustrated Guide to the Phyla of Life on Earth.* Freeman, San Francisco.

Parker S. P. (Ed.) (1982) *Synopsis and Classification of Living Organisms.* 2 Vols. McGraw-Hill, New York.

Phillipson J. (1981) Bioenergetic options and phylogeny. In *Physiological Ecology* (Ed.) Townsend C. R. & Calow P. pp. 20–45. Blackwell Scientific Publications, Oxford.

Sleigh M. A. (1979) Radiation of the eukaryote protista. In *The Origin of Major Invertebrate Groups* (Ed.) House M. R. pp. 23–53. Academic Press, London.

Whittaker R. H. (1969). New concepts of kingdoms of organisms. *Science, NY*, **163**, 150–160.

Monera

Buchanan R. E. & Gibbons N. E. (1974) *Bergey's Manual of Determinative Bacteriology*, 8th Edn. Williams & Wilkins, Baltimore.

Fox G. E. & 18 other authors (1980) The phylogeny of prokaryotes. *Science, NY*, **209**, 457–463.

Sieburth J. McN. (1979) *Sea Microbes.* Oxford University Press, New York.

Stanier R. Y., Adelberg E. A. & Ingraham J. L. (1977) *General Microbiology*, 4th Edn. Macmillan, London.

Protista

Bold H. C. & Wynne M. J. (1978) *Introduction to the Algae; Structure and Reproduction.* Prentice Hall, Englewood Cliffs.

Cox E. R. (1980) *Phytoflagellates.* Elsevier/North Holland, New York.

Dodge J. D. (1973) *The Fine Structure of Algal Cells.* Academic Press, London.

Levine N. D. *et al.* (1980) A newly revised classification of the Protozoa. *J. Protozool.*, **27**, 37–58.

Sleigh M. A. (1973) *The Biology of Protozoa.* Arnold, London.

Fungi

Ainsworth G. C. (1976) *Introduction to the History of Mycology.* Cambridge University Press, Cambridge.

Ainsworth G. C., Sparrow F. K. & Sussman A. S. (1973) *The Fungi: An Advanced Treatise.* Academic Press, New York.

Webster J. (1982) *An Introduction to the Fungi*. Cambridge University Press, Cambridge.

Plantae

Bell P. & Woodcock C. (1971) *The Diversity of Green Plants*, 2nd Edn. Arnold, London.

Bold H. C., Alexopoulos C. & Delevoras T. (1980) *Morphology of Plants and Fungi*, 4th Edn. Harper & Row, New York.

von Deuffer D., Schumacher W., Mägdefrau K. & Ehrendorfer F. (1976) *Strasburger's Textbook of Botany, New English Edn.* (Translation from 30th German Edn by Bell P. & Coombe D.) Longman, London.

Jones S. B. & Luchsinger A. E. (1979) *Plant Systematics*. McGraw-Hill, New York.

Animalia

Barnes R. D. (1980) *Invertebrate Zoology*, 4th Edn. Saunders, Philadelphia.

Boudreaux H. B. (1979) *Arthropod Phylogeny, with Special Reference to Insects*. Wiley, New York.

George J. D. & George J. J. (1979) *Marine Life*. Harrap, London.

House M. R. (Ed.) (1979) *The Origin of Major Invertebrate Groups*. Academic Press, London.

McFarland W. N., Pough F. H., Cade T. J. & Heiser J. B. (1979) *Vertebrate Life*. Macmillan, New York.

Index to Taxa

Note. Bacteria are entered under their common group name, not those of the formal taxonomic classes or orders to which they are sometimes apportioned.

Acantharea 55
Acanthobdellida 186
Acanthocephala 169
Acanthochitonida 189
Acanthopterygii 252
Acariformes 215
Acarpomyxea 51
Acipenserini 251
Acochlidioida 191
Acoela 150
Aconchulinida 55
Acrania (*see* Cephalochordata)
Acrasea (*see also* Acrasiomycetes) 52
Acrasida 52
Acrasiomycetes (*see also* Acrasea) 92
Acrothoracica 208
Acrotretida 199
Actinellida 56
Actiniaria 146
Actinobacteria 22
Actinomycetes (*see* Actinobacteria)
Actinomyxida 86
Actinophryida 58
Actinopoda 55
Actinosporea 86
Actinotrichida (*see* Acariformes)
Actinulida 142
Adenophorea 170
Aeolosomata 184
Aeolosomatida 185
Aerobic nitrogen-fixing bacteria 15
Afrenulata 179
Agamococcidiida 47
Agaricales 106
Agelasida 138
Agnatha 249
Air-tolerant anaerobic actinobacteria 23
Alcyonacea 145
Alcyonaria 144
Alismatanae 128
Amblypygi 214
Amiimorpha 251
Amoebiales 96
Amphibia 252
Amphidiscosida 134
Amphilinida 154
Amphinomida 182
Amphionidacea 211
Amphipoda 211
Anactinotrichida (*see* Parasitiformes)
Anaerobic gram-positive cocci 22
Anagalida 256
Anaspidacea 209
Anaxostylea 48
Andreaeanae 117
Angiospermae 119
Animalia 129
Annelida 180
Anomalodesmata 194
Anopla 161
Anoplura 230
Anostraca 204
Anoxyphotobacteria 14
Anthocerotales 116
Anthocerotopsida 116
Anthomedusae (*see* Athecata)
Anthozoa 143
Antipatharia 144
Anura 253
Aphasmidia (*see* Adenophorea)
Aphelenchida 173
Aphragmophora 233
Aphyllophorales 106
Apicomplexa 45
Aplotegmentaria 188

Aplousobranchia 246
Aplysiomorpha 191
Apodida 243
Aporida 156
Apostomatida 43
Appendaged bacteria (*see* Prosthecate bacteria)
Appendicularia (*see* Larvacea)
Arachnida 213
Araeolaimida 172
Aralianae 126
Araneae 214
Arbacioida 241
Arcellinida 51
Archaebacteria 11
Archaeogastropoda 190
Archaeognathata 226
Archegoniata 113
Archiacanthocephala 169
Archigregarinida 46
Archonta 257
Archoophora 151
Arcida 193
Arecanae 128
Arhynchobdellida 186
Arthracanthida 56
Arthrobacters 23
Arthropoda 203, 212, 221
Articulata 199
Ascaridida 173
Ascetospora 85
Ascidiacea 246
Ascomycotina 97
Ascothoracica 209
Asellariales 95
Aspidobothria 153
Aspidochirotida 243
Aspidogastrea 153
Aspiraculata 247
Asteranae 127
Asteroidea 238
Astomatida 43
Astrophorida (*see* Choristida)
Athalamida 54
Athecanephria 178
Athecata 141
Atherinomorpha 252
Auriculariales 107
Aves 254
Axinellida 137
Axostylea 49
Azygiida 154

Bacillariales 63
Bacillariophyceae 63
Bacillariophyta 63
Balanosporida 85
Bangiales 81
Bangiophyceae 81
Basidiomycotina 105
Basommatophora 192
Bathynellacea 210
Bdelloida 164
Bdelloidea 164
Bdellonemertea 162
Bdellovibrios 17
Beroida 149
Bicoecida (*see* Bicosoecida)
Bicosoecida 87
Biddulphiales 63
Biomyxida (*see* Athalamida)
Bivalvia 192
Bivalvulida 86
Blastocladiales (*see* Blastocladiida)
Blastocladiida 79
Blastocladiidea 79
Blastodiniales 34
Blastomycetes 103
Blattaria 229
Bodonida 35
Bourgueticrinida 237
Brachiopoda 198
Brachybasidiales 105
Branchiobdella 185
Branchiobdellida 186
Branchiopoda 204
Branchiura 208
Bromelianae 128
Bryophyta 114
Bryopsidales 77
Bryopsidophyceae 76
Bryozoa 200
Budding bacteria 18
Bullomorpha 190

Bursovaginida 160

Calanoida 207
Calcarea 135
Calobryales 115
Camallanida 173
Capitellida 182
Carnivora 257
Caryophyllanae 126
Caryophyllida 155
Cassiduloida 242
Catenulida 151
Caudata (*see* Urodela)
Caudofoveata 188
Celastranae 125
Centrales (*see* Biddulphiales)
Centrohelida 59
Cephalobaenida 220
Cephalocarida 204
Cephalochordata 248
Cephalodiscida 235
Cephalopoda 195
Ceramiales 82
Ceratoporellida 139
Ceriantharia 144
Ceriantipatharia 144
Cestida 149
Cestoda 155
Cestodaria 154
Chaetodermomorpha 188
Chaetognatha 233
Chaetonotida 163
Chaetophorales 76
Chamaesiphonales 12
Charales 75
Charophyceae 75
Chaunacanthida 56
Cheilostomata 201
Chelicerata 212
Chelonethida (*see* Pseudoscorpiones)
Chelonia 253
Chemolithotrophic bacteria 18
Chilopoda 224
Chlamydiales 19
Chlamydomyxales 69
Chlamydospermophyta (*see* Gnetopsida)
Chlorococcales 76
Chloromonadida (*see* Vacuolariales)
Chloromonadophyta (*see* Raphidophyta)
Chlorophyceae 75
Chlorophyta 74
Choanoflagellata 80
Chondrichthyes 249
Chondrophora 141
Chonotrichida 42
Chordata 245
Chordeumatida 223
Choristida 137
Chromadorida 172
Chroococcales 12
Chrysamoebidales 60
Chrysophyceae 60
Chrysophyta 60
Chrysosphaeriales 60
Chytridiales (*see* Chytridiida)
Chytridiida 78
Chytridiidea 78
Chytridiomycetes (*see also* Chytridiomycota) 93
Chytridiomycota (*see also* Chytridiomycetes) 78
Cidaroida 240
Ciliophora 40
Cirripedia 208
Cladocera 205
Cladocopida 205
Cladophorales 76
Clathrinida 135
Clitellata 185
Clupeomorpha 251
Clypeasteroida 242
Cnidaria 140
Coccidia 46
Coccobacilli 17
Coccosphaerales 71
Codiales 77
Coelacanthini 250
Coelomycetes 104
Coelotrophida 47
Coenothecalia 145
Coleoidea 195
Coleoptera 230
Collembola 226

Collothecida 165
Colourless sulphur bacteria 18
Colpodida 41
Comatulida 237
Commelinanae 128
Conchostraca 205
Coniferopsida 121
Conjugatophyceae 77
Copelata 247
Copepoda 206
Corallimorpharia 146
Cormophyta 113
Coronatae 143
Coronophorales 99
Corynebacteria 22
Cossurida 181
Craniata (*see* Vertebrata)
Crinoidea 237
Cristispires 20
Crocodilia 254
Crustacea 203
Cryptococcales (*see* Tetragonidales)
Cryptomonadales 73
Cryptonemiales 82
Cryptophyceae 73
Cryptophyta 73
Ctenodrilida 181
Ctenophora 148
Ctenostomata 201
Cubomedusae 143
Cubozoa 143
Cumacea 210
Cumulata (*see* Prolecithophora)
Curved bacteria 17
Cutleriales 64
Cyanobacteria 12
Cyanophyta (*see* Cyanobacteria)
Cycadales 123
Cycadopsida 122
Cyclophyllida 157
Cyclopoida 207
Cyclorhagida 166
Cyclostomata (Agnatha) 249
Cyclostomata (Bryozoa) (*see* Stenostomata)
Cydippida 148
Cyrtocrinida 237
Cyrtophorida 42
Cytophagas 21
Cyttariales 100

Dacrymycetales 105
Dactylochirotida 243
Dasycladales 77
Decapoda (Cephalopoda) (*see* Teuthoida)
Decapoda (Crustacea) 211
Demospongiae 136
Dendroceratida 139
Dendrochirotida 243
Dentalida 195
Dermaptera 228
Dermatophils 23
Derocheilocarida 206
Desmarestiales 64
Desmidiales 77
Desmodorida 172
Desmoscolecida 172
Desmothoracida 58
Deuteromycotina 103
Diadematoida 241
Diatomophyceae (*see* Bacillariophyceae)
Dibranchiata (*see* Coleoidea)
Dictyoceratida 139
Dictyochales 61
Dictyoptera (*see* Blattaria)
Dictyosteliia 52
Dictyosteliida 52
Dictyotales 65
Dicyemida 158
Digenea 153
Dillenianae 126
Dinoflagellata (*see* Dinophyta)
Dinophilida 184
Dinophyceae 33
Dinophysiales 33
Dinophyta 33
Diphyllida 156
Diplogasterida 173
Diplolepanae 117
Diplomonadida 48
Diplopoda 222
Diplura 225
Diplurata 225

Dipnoi 250
Diptera 232
Discomycetes 99
Dissimilatory sulphate-reducing bacteria 16
Doliolida 247
Dorylaimida 171
Dothideales 101

Ebriida 87
Eccrinales 95
Echinoderida (*see* Kinorhyncha)
Echinodermata 236
Echinoida 241
Echinoidea 240
Echinosteliida 53
Echinostomida 154
Echinothuroida 240
Echiura 176
Echiuroida 176
Ectocarpales 64
Ectoprocta (*see* Bryozoa)
Edentata 256
Elasipoda 243
Elopomorpha 251
Embiidida (*see* Embioptera)
Embioptera 228
Embryophyta 113
Endomycetales 97
Endospore-forming bacteria 21
Enopla 162
Enoplida 171
Enterobacteria 16
Enteropneusta 235
Entodiniomorphida 41
Entomophthorales 94
Entoprocta 202
Eoacanthocephala 169
Ephedrales 123
Ephemerida (*see* Ephemeroptera)
Ephemeroptera 227
Equisetales 120
Equisetopsida (*see* Sphenopsida)
Ericanae 126
Erysiphales 98
Euactinomycetes 23
Eubacteria 14
Euclasterida 239
Eucoccidiida 47
Euglenales 36
Euglenomorphales 37
Euglenophyceae 36
Euglenophyta 36
Eugregarinida 46
Eumycetozoa (*see* also Myxomycetes) 52
Eumycota 92
Eunicida 182
Euphausiacea 211
Eupodiscales (*see* Biddulphiales)
Eurotiales 98
Eustigmatales 70
Eustigmatophyta 70
Eutardigrada 219
Eutheria 256
Eutreptiales 36
Exobasidiales 105

Facultatively anaerobic rods 16
Filamentous gliding bacteria 21
Filicales 121
Filicopsida 120
Filosea 55
Filospermida 160
Flabelligerida 183
Florideophyceae 81
Flosculariida 165
Foraminiferida 54
Forcipulatida 239
Frenulata 178
Fucales 65
Fungi 89
Fungi Imperfecti (*see* Deuteromycotina)

Ganeshida 149
Gasteromycetes 107
Gastraxonacea 145
Gastropoda 189
Gastrotricha 163
Geophilida 225
Gigartinales 82
Ginkgoales 122
Ginkgoopsida 122

Gliding bacteria 20
Glomerida (*see* Oniscomorpha)
Glomeridesmida 222
Gnathostomula 160
Gnetales 123
Gnetopsida 123
Goniactinida 238
Gordioida 174
Gorgonacea 145
Gram-negative anaerobic fermenting bacteria 17
Gram-positive asporogenous bacteria 21
Granuloreticulosea 53
Green non-sulphur bacteria 14
Green sulphur bacteria 14
Gregarinia 46
Gromiida 55
Grylloblattaria 228
Gymnamoebida 50
Gymnodiniales 33
Gymnolaemata 201
Gymnophiona 253
Gymnosomata 191
Gymnospermae 119
Gymnostomatia 40
Gyrocotylida 154

Hadromerida 137
Halichondrida 138
Halobacteria 11
Halosphaerales 74
Hamamelidanae 125
Haplolepanae 117
Haplopharyngida 151
Haplosclerida 138
Haplotaxida 185
Haptophyceae (*see* Prymnesiophyceae)
Haptophyta 71
Harpacticoida 207
Harpellales 95
Helioporida (*see* Coenothecalia)
Heliozoea 58
Helotiales 99
Hemiascomycetes 97
Hemichordata 234
Hepaticopsida 114
Heterocyemida 158
Heteromyota 176
Heteronematales 37
Heteronemertea 162
Heteroptera 230
Heterotardigrada 218
Heterotrichida 44
Hexacorallia (*see* Zoantharia)
Hexactinellida 134
Hexactinosida 135
Hexapoda 225
Hirudinoida 186
Holasteroida 242
Holectypoida 242
Holobasidiomycetidae 105
Holocanthida 56
Holocephalii 250
Holocoela (*see* Prolecithophora)
Holothuroidea 243
Homalorhagida 167
Homoptera 230
Homosclerophorida 137
Hoplonemertea 162
Hydrozoa 141
Hydrurales 61
Hymenogastrales 107
Hymenomycetes 105
Hymenoptera 231
Hymenostomatia 43
Hymenostomatida 43
Hypermastigida 31
Hyphochytridiomycetes (*see also* Hyphochytridiomycotea) 93
Hyphochytridiomycotea (*see also* Hyphochytridiomycetes) 67
Hyphomycetes 103
Hypostomatia 42
Hypotrichida 44
Hysteriales 101

Impennes 255
Inarticulata 198
Inozoida 136
Insecta (*see* Hexapoda)
Insectivora 256
Ischnochitonida 189
Isochrysidales 71

Isocrinida 237
Isoetales 119
Isolaimida 171
Isopoda 211
Isoptera 229
Iuliformida 223

Juncanae 128
Jungermanniales 115

Karyorelictida 40
Kinetofragminophorea 40
Kinetoplasta 35
Kinorhyncha 166

Laboulbeniales 102
Laboulbeniomycetes 102
Labyrinthomorpha 68
Labyrinthulea 68
Lactobacteria 22
Lagenidiales (*see* Lagenidiida)
Lagenidiida 67
Laingiomedusae 142
Lamellibranchia (*see* Bivalvia)
Lamianae 126
Laminariales 65
Larvacea 247
Lecanicephalida 155
Lecanorales 100
Lecithoepitheliata 151
Lecithophora (*see* Neorhabdocoela)
Lepidopleurida 189
Lepidoptera 232
Leptomedusae (*see* Thecata)
Leptomitales (*see* Leptomitida)
Leptomitida 67
Leptomyxida 51
Leptospires 20
Leptostraca 209
Leucettida 135
Leucosoleniida 136
Liceida 53
Lilianae 128
Liliopsida 127
Limacomorpha (*see* Glomeridesmida)
Limnomedusae 142
Linguatulida (*see* Pentastoma)
Lingulida 199
Lithistida 137
Lithobiida 224
Litobothrida 156
Lobata 149
Lobosea 50
Loculoascomycetes 100
Lumbriculida 185
Lychniscosida 135
Lycoperdales 108
Lycopodiales 119
Lycopsida 119
Lyssacinosida 135

Macrodasyida 163
Macrostomida 151
Madreporaria (*see* Scleractinia)
Magnolianae 124
Magnoliopsida 124
Malacostraca 209
Mallophaga 229
Malvanae 126
Mammalia 255
Mantodea 229
Marattiales 120
Marchantiales 115
Marsileales 121
Marsupialia 256
Mastigomycotina (*see also* Phycomycota) 92
Mecoptera 231
Medusozoa 140
Megaloptera 231
Melanconiales 104
Melanogastrales 108
Meliolales 99
Mermithida 171
Merostomata 212
Mesogastropoda 190
Mesotardigrada 219
Mesozoa 158
Metamonada 48
Metazoa 131
Metchnikovellida 83

Methanogenic bacteria 11
Methylomonads 19
Metzgeriales 115
Micrococci 22
Microcoryphia 226
Microspora 83
Microsporea 83
Microsporida 84
Microthyriales 101
Milleporina 141
Millericrinida 237
Minisporida 83
Mischococcales 69
Misophrioida 207
Mollicutes (*see* Mycoplasms)
Mollusca 187
Molpadiida 244
Monera 7
Monhysterida 172
Moniligastrida 185
Monoblepharidales (*see* Monoblepharidiida)
Monoblepharidiida 79
Monocleales 115
Monogenea 152
Monogonata 165
Mononchida 171
Monopisthocotylea 153
Monoplacophora 188
Monothalamida 54
Monstrilloida 207
Mormonilloida 207
Mucorales 94
Multivalvulida 86
Muscopsida 116
Muspiceida 171
Mycoplasms 23
Myida 194
Myodocopida 205
Myriangiales 101
Myriapoda 222
Myrientomata 226
Myrtanae 125
Mysidacea 210
Mystacocarida 206
Mytilida 193
Myxobacteria 20
Myxogastria 53
Myxomycetes (*see also* Eumycetozoa) 92
Myxomycota 92
Myxosporea 86
Myxozoa 86
Myzostomida 184

Narcomedusae 142
Nassellarida 57
Nassulida 42
Nautiloida 195
Nautiloidea 195
Nectonematida 174
Nemalionales 82
Nemata (*see* Nematoda)
Nematoda 170
Nematomorpha 174
Nematophora (*see* Chordeumatida)
Nemertea 161
Nemertini (*see* Nemertea)
Neogastropoda 190
Neognathae 255
Neogregarinida 46
Neolampadoida 242
Neomeniomorpha 188
Neorhabdocoela 152
Neoselachii 249
Nerillida 184
Neuroptera 231
Nidulariales 108
Nippotaeniida 156
Nitrifying bacteria 18
Noctilucales 34
Nostocales 13
Notandropora (*see* Catenulida)
Notaspida (*see* Pleurobranchomorpha)
Notodelphyoida 207
Notostigmata 215
Notostraca 205
Nuculida 193
Nuda 149
Nudibranchia 191
Nymphaeanae 125

Ochromonadales 60
Octocorallia (*see* Alcyonaria)
Octopoda 196

Odonata 227
Odontostomatida 44
Oedogoniales 76
Oedogoniophyceae 76
Oegophiurida 239
Oligochaeta 185
Oligoentomata 226
Oligohymenophorea 43
Oligotrichida 44
Onchidida 192
Oniscomorpha 222
Onychophora 217
Oomycetes (*see also* Oomycotea) 93
Oomycotea 66
Opalinata 38
Opheliida 182
Ophioglossales 120
Ophiurida 240
Ophiuroidea 239
Opiliocariformes (*see* Notostigmata)
Opiliones 215
Opisthandropora (*see* Macrostomida)
Opisthobranchia 190
Opisthorchiida 154
Orbiniida 181
Ornithodelphia 256
Orthonectida 159
Orthoptera 228
Ostariophysi 252
Osteichthyes 250
Osteoglossomorpha 251
Ostracoda 205
Ostropales 100
Oweniida 183
Oxymonadida 49
Oxyphotobacteria 12

Pachytegmentaria 188
Palaeacanthocephala 169
Palaeocopida 206
Palaeognathae 255
Palaeonemertea 161
Palmariales 82
Palpigradi 214
Pantopoda (*see* Pycnogonida)
Parabasalia 31
Paracanthopterygii 252
Paramyxea 85
Paramyxida 85
Parasitiformes 215
Pauropoda 224
Pavlovales 72
Paxillosida 238
Pedinellales 61
Pedinoida 241
Pedinomonadales 74
Pelecypoda (*see* Bivalvia)
Pelobiontea 51
Pennales (*see* Bacillariales)
Pennatulacea 145
Pentastoma 220
Peridiniales 34
Peritrichia 44
Peritrichida 44
Perkinsea 45
Peronosporales (*see* Peronosporida)
Peronosporida 67
Petrosiida 138
Pezizales 99
Phacidiales 100
Phaeocalpida 57
Phaeoconchida 58
Phaeocystida 57
Phaeodarea 57
Phaeodendrida 58
Phaeogromida 57
Phaeophyceae 64
Phaeophyta 64
Phaeosphaerida 57
Phaeothamniales 60
Phallales 107
Phasmida 228
Phasmidia (*see* Secernentea)
Phlebobranchia 246
Pholidota 256
Phorona 197
Phoronida 197
Phragmobasidiomycetidae 106
Phragmophora 233
Phrynophiurida 240
Phycomycota (*see also* Mastigomycotina) 66
Phylactolaemata 200
Phyllodocida 182
Phymosomatoida 241

Physarida 53
Phytodinales 34
Pinales 122
Pinopsida (*see* Coniferopsida)
Piroplasmia 47
Placozoa 133
Plagiorchiida 154
Planctosphaerida 235
Plantae 111
Plasmodiophorea (*see also* Plasmodiophoromycetes) 54
Plasmodiophoromycetes (*see also* Plasmodiophorea) 92
Platyasterida 238
Platycopida 206
Platyctida 149
Platyhelminthes 150
Plecoptera 228
Plectomycetes 98
Pleosporales 101
Pleurobranchomorpha 191
Pleurocapsales 13
Pleurogona 246
Pleurostomatida 41
Ploima 165
Plumatellida 201
Podaxales 107
Podocopida 206
Poecilosclerida 138
Poecilostomatoida 208
Poeobiida 183
Pogonophora 178
Polychaeta 181
Polycladida 152
Polycystinea 56
Polydesmida 224
Polygordiida 184
Polyhymenophorea 44
Polyopisthocotylea 153
Polyplacophora 189
Polypterini 250
Polytrichanae 117
Polyxenida 222
Polyzoa (*see* Bryozoa)
Polyzoniida 223
Porifera 134
Porocephalida 220
Porphyridiales 81
Prasinocladiales 75
Prasinophyceae 74
Prasiolales 75
Priapula 168
Priapulida 168
Proactinomycetes 22
Prochlorophytes 13
Prokaryota (*see* Monera)
Prolecithophora 151
Prorocentrales 33
Proseriata 152
Prosobranchia 190
Prosthecate bacteria 18
Prostomatida 41
Protacanthopterygii 252
Proteanae 126
Proteocephalida 156
Proteromonadida 87
Protista 25
Protoalcyonaria 144
Protoctista (*see* Protista)
Protodrilida 184
Protomycetales 97
Protosteliia 52
Protosteliida 52
Protozoa 131
Protura 226
Prymnesiales 71
Prymnesiophyceae 71
Prymnesiophyta (*see* Haptophyta)
Psamminida 54
Psammodrilida 181
Pseudomonads 15
Pseudophyllida 155
Pseudoscorpiones 214
Psilotales 119
Psilotopsida 119
Psocoptera 229
Pteridophyta 118
Pteriida 193
Pterobranchia 235
Pterospermatales 74
Pterygota 227
Ptychodactiaria 147
Pulmonata 192
Purple non-sulphur bacteria 15
Purple sulphur bacteria 15
Pycnogonida 215

Pyramidellomorpha 190
Pyramimonadales 75
Pyrenomycetes 98
Pyrosomida 247

Questida 182

Ranunculanae 125
Raphidioida 231
Raphidophyta 62
Remipedia 204
Reptilia 253
Retortamonadida 48
Rhabditida 172
Rhabdomonadales 36
Rhabdopleurida 235
Rhizocephala 209
Rhizopoda 50
Rhizostomeae 143
Rhodophyta 81
Rhodymeniales 82
Rhombozoa 158
Rhynchobdellida 186
Rhynchocephalia 254
Rhynchocoela (*see* Nemertea)
Rhynchodida 42
Rhynchonellida 199
Ricinulei 214
Rickettsiales 19
Rickettsias 19
Rodentia 257
Rosanae 125
Rotifera 164
Rudimicrosporea 83
Rutanae 125

Sabellida 183
Sacoglossa 191
Salenoida 241
Salpida 247
Salviniales 121
Saprolegniales (*see* Saprolegniida)
Saprolegniida 67
Sarcodina 50
Scaphopoda 194
Schizomida 213
Schizopyrenida 51
Scleractinia 146
Sclerodermatales 108
Sclerospongiae 139
Scolopendrida 225
Scorpiones 213
Scuticociliatida 43
Scutigerida 224
Scyphozoa 142
Secernentea 172
Seisonidea 164
Selaginellales 120
Semaeostomeae 143
Semionotini 251
Sepioida 196
Septibranchia 194
Septobasidiales 107
Seticoronaria 168
Sheathed bacteria 17
Siderocapsans 18
Silicoflagellida (*see* Dictyochales)
Siphonaptera 232
Siphonodentalida 195
Siphonophora 142
Siphonostomatoida 207
Siphunculata (*see* Anoplura)
Sipuncula 175
Solemyida 193
Solenogastres (*see* Neomeniomorpha)
Solpugida 215
Somasteroidea 238
Spatangoida 242
Spathebothriida 155
Spelaeogriphacea 210
Spermatophyta 118
Sphacelariales 65
Sphaeriales 99
Sphaerocarpales 116
Sphaeropsidales 104
Sphagnanae 116
Sphenomonadales 37
Sphenopsida 120
Sphinctozoida 136
Spintherida 182
Spinulosida 239
Spionida 181
Spirillas 16

Spirobolida 223
Spirochaetales 20
Spirochaetes 19
Spirophorida 137
Spirurida 173
Spizellomycetales (*see* Spizellomycetida)
Spizellomycetida 79
Sporozoea 45
Spumellarida 57
Squamata 254
Stannominida 54
Stauromedusae 143
Stellatosporea 85
Stemmiulida 223
Stemonitida 53
Stenoglossa (*see* Neogastropoda)
Stenolaemata 201
Stenostomata 201
Stephanopogonomorpha 39
Stereomyxida 52
Sternaspida 183
Stigonematales 13
Stolidobranchia (*see* Pleurogona)
Stolonifera 145
Stomatopoda 209
Strepsiptera 230
Strigeidida 153
Strongylida 173
Stygocaridacea 210
Stylasterina 141
Stylommatophora 192
Suctorida 43
Sulphate-reducing bacteria (*see* Dissimilatory sulphate-reducing bacteria)
Sycettida 136
Symphyacanthida 56
Symphyla 225
Syndiniales 34
Synhymenida 42
Systellommatophora (*see* Onchidida)

Tabulospongida 139
Taeniida (*see* Cyclophyllida)
Taenioglossa (*see* Mesogastropoda)
Tanaidacea 210
Taphrinales 97
Tardigrada 218
Taxales 122
Taxopodida 58
Telestacea 145
Teliomycetes 108
Temnocephalida 152
Temnopleuroida 241
Tentaculata 148
Terebellida 183
Terebratulida 199
Testudines (*see* Chelonia)
Tetrabranchiata (*see* Nautiloidea)
Tetragonidales 73
Tetraphidanae 117
Tetraphyllida 156
Tetrarhynchida (*see* Trepanorhynchida)
Tetrasporales 75
Teuthoida 196
Thalassocalycida 149
Thaliacea 247
Thallochrysidales 61
Thecanephria 179
Thecata 142
Thecideidida 199
Thecosomata 191
Thermoacidophilic bacteria 11
Thermosbaenacea 210
Thoracica 208
Thraustochytridea 68
Thysanoptera 230
Thysanura 227
Tracheophyta 118
Trachylina 142
Trachymedusae 142
Trematoda 153
Tremellales 106
Trepanorhynchida 155
Tribonematales 69
Tribophyceae (*see* Xanthophyceae)
Trichiida 53
Trichocephalida 171
Trichomonadida 31
Trichomycetes 95
Trichoptera 232
Trichosida 51
Trichostomatida 41

Trichosyringida (*see* Mermithida)
Tricladida 152
Trigoniida 193
Tryblidioida 188
Trypanosomatida 35
Tuberales 100
Tulasnellales 106
Tulostomatales 108
Tunicata (*see* Urochordata)
Turbellaria 150
Tylenchida 173
Typhlogena 223

Ulotrichales 76
Ulvales 76
Ungulata 257
Unionida 194
Uniramia 221
Uredinales 109
Urochordata 245
Urodela 253
Uropygi 213
Ustilaginales 109

Vacuolariales 62
Valvatida 238
Vampyromorpha 196
Vaucheriales 69
Venerida 194
Verongiida 139
Vertebrata 248
Vestibuliferia 41
Vestimentifera 179
Vibrios 16
Volvocales 75

Welwitschiales 124

Xanthophyceae 69
Xanthophyta 69
Xenophyophorea 54
Xenopneusta 177
Xiphosura 213

Zoantharia 146
Zoanthidea (*see* Zoanthinaria)
Zoanthinaria 146
Zoopagales 95
Zoraptera 229
Zygnemaphyceae (*see* Conjugatophyceae)
Zygnematales 77
Zygoentomata 227
Zygomycetes 94
Zygomycotina 94